TriMathlon

TriMathlon
A Workout Beyond
the School Curriculum

Judith D. Sally
Northwestern University

Paul J. Sally, Jr.
University of Chicago

A K Peters
Natick, Massachusetts

Editorial, Sales, and Customer Service Office

A K Peters, Ltd.
63 South Avenue
Natick, MA 01760
www.akpeters.com

Library of Congress Cataloging-in-Publication Data

Sally, Judith D
  TriMathlon : a workout beyond the school curriculum / Judith D. Sally, paul J. Sally, Jr.

    p. cm.

  Includes index
  ISBN 1-56881-184-5
   1. Number theory-Problems, exercises, etc. 2. Geometry-Problems, exercises, etc. I. Sally, Paul. II. Title.

QA241. S22 2003
512'.7'06–dc21

2002193155

Printed in the United States of America
06 05 04 03 02   10 9 8 7 6 5 4 3 2 1

To Our Grandchildren:
(in order of appearance)

Ben

Chris

Tim

Mike

Sean

Tom

Colleen

and

Rachel

Contents

To the Player   ix

To the Coach   xi

Acknowledgments   xiii

Part I   Swim: Arithmetic   1

1   Race to 100   3

2   Roll Back   9

3   Wordsworth   15

Part II   Bike: Numbers and Symmetry   23

4   The Triangle and Square Games   25

5   Palindromes   53

6   The Four Numbers Game   77

Contents

Part III   Run: Geometry   **111**

7   Tessellation   113

8   Circle Packing in the Plane   145

9   Lattice Polygons   171

10 Dissection   203

To the Player

In *TriMathlon: A Workout Beyond the School Curriculum* you will find games, problems and investigations designed to flex your math muscles and give you a new perspective on mathematics. The guided activities are fun, interesting and challenging—you will be introduced to some truly heavy-weight mathematical ideas. The strenuous mental activity often required has as its reward the satisfaction and confidence that accompany meaningful investigations of mathematical ideas.

TriMathlon has three parts: Swim, Bike, and Run. *Swim* contains warm-up games in arithmetic; *Bike* consists of more ambitious projects in numbers and symmetry; and *Run* includes challenging workouts in geometry. All three parts start at an easy pace in familiar territory. Many of the chapters contain a section called Heavy Lifting ◁△▷ for those who want to press to the max!

All information needed to solve the problems (as well as hints and suggestions) is provided, but you are to work out the solutions on your own or with a group of your classmates. Throughout, there are questions followed by a stopwatch ⏱ or a triathlete ⚓, 🚴, 🏃. These icons mean "Pause; take time to think and to work on your own." A pencil in the margin means that you should fill in the accompanying table.

Solutions to all of the numbered questions, challenges, etc. in the text are found in separate solutions sections.

Now it's time to see how far *TriMathlon* can take you. Go for it.

To the Coach

TriMathlon: A Workout Beyond the School Curriculum invites young students to participate in challenging excursions into areas of number theory and geometry that extend beyond the borders of the basic mathematics in the school curriculum. The activities in this book present students with a dynamic and fresh perspective on mathematics, and encourage active and energetic response. Some of the projects in this book are aimed at inquisitive fourth and fifth graders, others are more appropriate for students in the sixth through eighth grades. Suggestions for further projects accompany each activity. References also include web resources.

TriMathlon contains three parts consisting of ten guided activities that are appropriate for students working on their own or in small groups. Part I consists of three warm-up games suitable for any age level. These games require some thinking, some ingenuity, and even some non-trivial mathematical vocabulary.

Part II has three more challenging activities. They lead to ideas that can be projected into quite sophisticated mathematics. Nonetheless, these problems in numbers and symmetry can be understood and attacked by students in the lower and middle grades.

Finally, in Part III, students are engaged in interesting and provocative notions about geometric objects. The four problems in Part III represent a level of thinking that carries over into college and university mathematics.

The initial concepts in each of the ten activities are presented in such a way that enthusiastic students can work successfully at the start of each one. In Parts II and III, the investigations proceed from familiar areas to new locales and finally in the special Heavy Lifting section, the students are really put through their paces. The idea being conveyed to the students is that mathematics does not occur in discrete or isolated blocks, but is a subject that grows from and expands on concepts and results that have been obtained previously. Mathematics is not simply a matter of memorizing and applying formulas. Students will find that many problems cannot be solved at first glance, but require careful thought and attention over a period of time before complete understanding is reached. Sometimes students have the incorrect idea that if they cannot solve a problem, it is because they have forgotten a formula they are supposed to know. You can encourage them to rely on their own creativity and ingenuity to find a solution. With your help, the projects in this book will increase the mathematical understanding and confidence of the students.

TriMathlon introduces the mathematical ideas involved in each activity, describes the mathematical goals, and acts as a guide and coach along the way. Throughout, there are questions followed by a stopwatch 🕭 or a triathlete ⮿, 🏊, 🏃. These icons mean "Pause; take time to think and to work on your own." We ask that you remind the students to use the these icons as intended. A pencil in the margin means that the student should fill in the accompanying table. Your guidance, suggestions and support will be of great value. In no case, however, should the solution be provided. Students are to work out solutions for themselves. Solutions to all of the numbered questions, challenges, etc. in the text are found in separate solutions sections.

Acknowledgments

Our fascination with mathematical problems that begin with very elementary ideas and grow into broader and more sophisticated theories span many decades.

Race to 100 and Roll Back are variations of well-known mathematical games. A puzzle described by Puzzlemaster Will Shortz on NPR evolved into Wordsworth. The Triangle and Square Games were shown to us by Barry Cipra. Palindromes were part of our earliest contact with mathematics. We first encountered The Four Numbers Game in articles in *The American Mathematical Monthly, The Fibonacci Quarterly and Mathematics Magazine.* Our interest in the ancient subject of tessellation was sparked many years ago by discussions with Sam Savage and Doris Schattschneider. We learned much about packing from Greg Kupferberg. Problems in lattice polygons came to our attention in the thesis of Jamie Pommerscheim. Finally, we have been intrigued by dissection ever since we learned about the Banach-Tarski paradox in graduate school.

The problems in *TriMathlon* have been used in various forms in "Take Our Daughters to Work Day" workshops at Northwestern University, in the University of Chicago Young Scholars Program for seventh through twelfth graders, with middle grade teachers from the Chicago Public Schools in the University of Chicago SESAME staff and development program, and in our undergraduate classes.

We are greatful to many people for their help with this book. First, we thank Bill Casselman and Naomi Fisher for helpful comments about Chapter 10. We also thank Barbara Csima, Cameron Freer, Katie O'Shea, Nick Ramsey, Ashley Reiter and Shaffiq Welji for reading and commenting on various versions of the problems in the manuscript. Special thanks are due to Grace Jang and Bridget Tenner who read and commented on the entire manuscript. We express our sincere gratitude to Charlotte Ochanine who did a masterful job T_EXing the manuscript. Finally, we thank Alice and Klaus Peters and the staff of A K Peters, Ltd. for their excellent work producing TriMathlon.

Judith D. Sally
Paul J. Sally, Jr.
September 30th, 2002

Swim

Arithmetic

Chapter One
Race to 100

In this simple game of addition, use your ingenuity to find a strategy to be the first to reach 100.

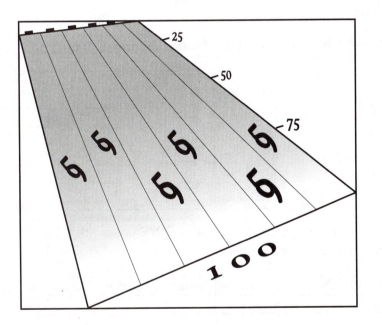

"O my friends, what are those wonderful numbers about which you are reasoning?..."

Plato

Here is an interesting mathematical game that can be played with a friend anytime, anyplace. All you need to know is how to add. The challenge is trying to discover the strategy for winning every game.

The rules of the game are very easy. In the first round, Player 1 picks a number between 1 and 10. Then Player 2 also picks a number between 1 and 10, and adds it to the number chosen by Player 1. For example, if I start first and pick 4, you might pick 7 and add it to 4 to make 11. In the following rounds, the players continue in turn, picking a number, always between 1 and 10, and adding it to the previous sum. The game ends when one player, the winner, hits 100. That's all there is to it!

Here is a sample game. Say I start first and pick 6, and then you pick 8, so the sum at the end of the first round is 14. In the second round, if I pick 10, the sum would be 24, and if you also pick 10, the sum would be 34. Here are the remaining rounds of this game between you and me. You fill in the sums.

Round	Number I Pick	Sum	Number You Pick	Sum
1	6	6	8	14
2	10	24	10	34
3	2	____	7	____
4	9	____	4	____
5	1	____	10	____
6	8	____	6	____
7	2	____	6	____

There is no winner yet, but I can see that, if you are alert, I am going to lose the game. No matter what number I pick in the next round, you can win.

Round	Number I Pick	Sum	Number You Pick	Sum
8	8			100

Now, find another partner and play the game a few times before reading on.

Some games are determined purely by chance, but this game has a winning strategy. If you go first, there is a way for you to win no matter which numbers your opponent chooses. Let's figure out what that strategy is.

To do this we work backward from 100. (Sometimes it pays to go backward!) You want to be the one to hit 100, so you need your opponent to give you a sum of at least 90. This means that if you can get to exactly 89 on the turn before, you can win (no larger than 89, of course, or you might lose). How do you get to 89? You need your opponent to give you a sum of at least 79. So if you can reach exactly 78, you can win. To get to 78, you need your opponent to give you a sum of at least 68. So if you can reach 67, you can win. Thus, the target numbers, that is, the numbers to shoot for, are 100 to win, before that 89, before that 78, and before that 67.

Now you probably see what the pattern is, so let's make a chart from the last round going backward to the first round. Since we don't know what numbers your opponent will pick, the chart shows the sums you need to aim for and the range of where the sum will be after your opponent's turn.

Round	Sum after Your Turn	Sum after Your Opponent's Turn
last	100 (winner!!!)	
second to last	89	$90 \leq$ sum ≤ 99
	78	$79 \leq$ sum ≤ 88
	67	$68 \leq$ sum ≤ 77
	56	$57 \leq$ sum ≤ 66
	45	$46 \leq$ sum ≤ 55
fourth	34	$35 \leq$ sum ≤ 44
third	23	$24 \leq$ sum ≤ 33
second	12	$13 \leq$ sum ≤ 22
first	1	$2 \leq$ sum ≤ 11

Did you figure out the winning strategy?

> The strategy is to be the first player
> and to pick the number 1.

So if you take the first turn and pick the number 1, then no matter which numbers your opponent chooses, you win if you follow the plan.

Now try the sample game below. You start first. See if you can make the choices needed to follow the strategy based on your opponent's choices given below. Remember that you want to pick the number 1 to start, and then hit each of the target numbers until you hit 100 to win.

Sample Game 1

Round	You Pick	Sum	Your Opponent Picks	Sum
first	____	1	9	10
second	____	12	7	19
third	____	23	10	33
fourth	____	34	1	35
fifth	____	45	5	50
sixth	____	56	2	58
seventh	____	67	9	76
eighth	____	78	10	88
ninth	____	89	3	92
tenth	____	100		

What can you do if your opponent takes the first turn? If your opponent has figured out the strategy and carries it out, you lose. If not, then on some turn your opponent might not make the correct choice to get one of the target numbers $1, 12, 23, 34, 45, 56, 67, 78,$ or 89, needed to win. On your turn, you can add the difference to hit one of the targets exactly and, once again, you can win.

Here are the first few turns of a sample game where your opponent starts first. We will assume that your opponent does not know the strategy. See if you can win.

Sample Game 2

Round	Your Opponent Picks	Sum	You Pick	Sum
first	1	1	5	6
second	6	12	5	17
third	5	22	____	____

It does not matter what number you pick for your first two turns because your opponent has, by chance, picked numbers to get the sums 1 and 12, which you know are the ones needed to assure a win. But in the third turn, the sum after your opponent's pick is only 22, so what number should you pick? Can you pick a number that will guarantee a win for you if you follow the plan?

Strategy is an important part of many games and many mathematical problems. Now play the Race to 100 Game with some friends. See if they realize that there is a winning strategy.

Solutions

Sample Game 1. If you chose the numbers $1, 2, 4, 1, 10, 6, 9, 2, 1, 8$ you followed the plan perfectly and are ready to win a real game (against someone who has not yet figured out the winning strategy, of course!).

Sample Game 2. If you picked the number 1, you are right.

Suggestions for the Endurance Athlete

10K Challenge. Let's keep the basic idea of the game the same but increase the numbers. Can you figure out a winning strategy for the Race to 1000 Game, where each player is allowed to pick a number between 1 and 25?

20K Challenge. Here's a variation of the Race to 100 Game where the range of allowable numbers to pick increases with each round. In the first round, each player must choose a number between 1 and 10. In the second round, each player is allowed to choose a number between 1 and 11. In the third round, number choices are allowed between 1 and 12, etc. Can you figure out a winning strategy? What happens if you decrease the range of choices with each round? Is there a winning strategy?

Chapter Two

Roll Back

This is a game of chance. Try your luck, but don't roll into negative number territory.

"... somewhere between chance and mystery lies imagination..."

L.Buñuel

Since the Race to 100 Game was based on addition of numbers, how about a game that uses subtraction? This game is a game of chance. It can be played on your own or with friends.

You will need a die, that is, one of a pair of dice, or, if you don't have a die, make yourself a spinner with the numbers from 1 to 6 on it. The game of Roll Back works like this. Starting with the number 100, subtract numbers determined by rolls of the die. The objective is to roll back as close as you can to 0 with no negative numbers allowed.

Roll Back Rules (*for one player*): You must roll the die a total of five times. To start, write down the number 100, and take your first roll of the die. You now have two choices: Either subtract the number rolled (that is, the number on top of the die) from 100 and write down the result, or multiply the number rolled by 10 and then subtract it from 100 and write that result down. So if you roll a 4, you can roll back to $100 - 4 = 96$ or to $100 - 40 = 60$.

You now have four more rolls of the die and the same two choices on each roll. The goal is to roll back to 0, or as close to 0 as you can, without ever going below 0 into negative number territory. You must use all five rolls of the die and, if you go below 0, you lose.

Here is a sample game.

Roll	Number Rolled	Choice	Result
first	4	4×10	$100 - 40 = 60$
	2	2×10	$60 - 20 = 40$
	4	4 *	$40 - 4 = 36$
	1	1×10	$36 - 10 = 26$
fifth	6	6 *	$26 - 6 = 20$

* There is no other possible choice here. Why?

20 is not very close to 0. Do you think that if we had made different choices, we would have gotten closer to 0?

In the next example, the rolls of the die are the same, but different choices allow us to roll back closer to 0.

Roll	Number Rolled	Choice	Result
first	4	4×10	$100 - 40 = 60$
	2	2	$60 - 2 = 58$
	4	4×10	$58 - 40 = 18$
	1	1×10	$18 - 10 = 8$
fifth	6	6	$8 - 6 = 2$

You can check all the other possible choices to see that 2 is the best possible result with the given rolls of the die.

Now it's time for you to roll the die and play the game.

Here are variations of the game for more than one player.

Roll Back for Several Players. Each participant uses a separate die and plays the game following the rules above, with 5 rolls for each player. The winner is the player who rolls back closest to 0, with no negative numbers allowed.

Roll Back for Two Players. The players share the die, share the 100 and play off one another's result. Turns alternate and the players roll, choose, and subtract as before but this time there is no limit on the number of rolls and there are two ways to win. You win if you roll back exactly to 0, or if your opponent is forced to go below 0.

Roll Back Invented by You. There are many variations of Roll Back. Why don't you create one?

After you have played a few rounds of Roll Back, think about the following questions.

Question 1. Is it true that multiplying the first roll by 10 is always the best choice? Try to make up a game where choosing "first roll × 10" is not the best strategy.

Question 2. Suppose, as in the following game, the sum of all the numbers rolled is exactly 10.

Roll	Number Rolled	Choice	Result
first	3	3×10	$100 - 30 = 70$
	1	1×10	$70 - 10 = 60$
	4	4×10	$60 - 40 = 20$
	1	1×10	$20 - 10 = 10$
fifth	1	1×10	$10 - 10 = 0$

If you make the choice of multiplying each number rolled by 10, as I did above, will you always roll back to 0? Why?

Of course, you won't know when you are playing what the sum of all the rolls will be. You have to make your choices in the game **before** you know what all the rolls are. So you have to develop a strategy for the game that only depends on the previous rolls of the die.

Remember, this is a game of chance and the roll of the die is unpredictable.

Solutions

Question 1. Here is a game that shows multiplying by 10 on the first roll might not roll you back closest to 0.

Roll	Number Rolled	Choice	Result
first	6	6×10	$100 - 60 = 40$
	6	6 *	$40 - 6 = 34$
	4	4 *	$34 - 4 = 30$
	4	4 *	$30 - 4 = 26$
fifth	3	3 *	$26 - 3 = 23$

* No other allowable option here.

Suppose, instead, different choices are made, for example

Roll	Number Rolled	Choice	Result
first	6	6	$100 - 6 = 94$
	6	6	$94 - 6 = 88$
	4	4×10	$88 - 40 = 48$
	4	4×10	$48 - 40 = 8$
fifth	3	3	$8 - 3 = 5$

The end result, 5, is closer to 0 than 23.

Question 2. The solution uses the distributive law of arithmetic. Suppose the numbers rolled are r_1, r_2, r_3, r_4, r_5 and that

$$r_1 + r_2 + r_3 + r_4 + r_5 = 10.$$

Then, by the distributive law,

$$100 - (r_1 \times 10) - (r_2 \times 10) - (r_3 \times 10) - (r_4 \times 10) - (r_5 \times 10) =$$
$$100 - ((r_1 + r_2 + r_3 + r_4 + r_5) \times 10).$$

But $r_1 + r_2 + r_3 + r_4 + r_5 = 10$, so

$$100 - ((r_1 + r_2 + r_3 + r_4 + r_5) \times 10) = 100 - (10 \times 10)$$
$$= 100 - 100$$
$$= 0.$$

Consequently, if, by chance, $r_1 + r_2 + r_3 + r_4 + r_5 = 10$, and if you multiply by 10 at each turn, you can always roll back to 0.

Suggestions for the Endurance Athlete

10K Challenge. In answering Question 1, you saw that it is not always the best strategy to multiply the first roll of the die by 10. You also saw in Question 2 that, if very special circumstances arise by chance, then multiplying each roll by 10 is the best strategy. Can you figure out a good strategy for the first roll of the game? For example, on the first roll, is there a cutoff number C so that if your first roll $r_1 \leq C$, then the best strategy is to multiply r_1 by 10 and otherwise, not?

Chapter Three

Wordsworth

The title of this game is a play on words. You may recognize the name Wordsworth as the surname of the late nineteenth century English poet, William Wordsworth. The game Wordsworth gives a number value, or worth, to words by applying the mathematical operations of addition and multiplication. Mathematics and a dictionary are the tools needed for this game.

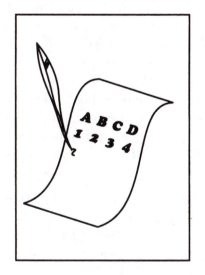

"At evening when with pencil and with slate....
With crosses and with cyphers scribbled o'er,
We schemed and puzzled head opposed to head."
W. Wordsworth

Who said you can't add apples and oranges? Here's a fun and easy way to do it. Give each letter of the alphabet, starting with A, a number value from 1 to 26. So A has value 1, B has value 2, etc., as in the following table.

A	B	C	D	E	F	G	H	I	J	K	L	M
1	2	3	4	5	6	7	8	9	10	11	12	13

N	O	P	Q	R	S	T	U	V	W	X	Y	Z
14	15	16	17	18	19	20	21	22	23	24	25	26

Now, give a word, such as "apples," a number value equal to the sum of the values of its letters. This means "apples" is worth $1+16+16+12+5+19 = 69$. "Oranges" is worth $15 + 18 + 1 + 14 + 7 + 5 + 19 = 79$. So just for fun, we can "add" apples and oranges to get $69 + 79 = 148$.

You can think of the letter values as money values if you want. As you are undoubtedly aware, the business world is very conscious of the monetary value of certain words and names. On our scale, in terms of dollars (you could use francs, pesos, or whatever), Michael Jordan's name is worth $\$13 + \$9 + \$3 + \$8 + \$1 + \$5 + \$12 + \$10 + \$15 + \$18 + \$4 + \$1 + \$14 = \$113$ and Serena Williams' name is worth $\$19 + \$5 + \$18 + \$5 + \$14 + \$1 + \$23 + \$9 + \$12 + \$12 + \$9 + \$1 + \$13 + \$19 = \$160$. How much is your name worth?

Let's make this game more interesting by reversing the order of what we have been doing. This means we are going to first pick a number, and then look for common English words whose value is equal to that number. No proper names are allowed this time, but you can use a dictionary anytime you want.

Question 1. Suppose the number picked is 23. Here are some words with value 23: BAT, ELF, DEN. What are some other words with value 23? What is the longest word you can find with value 23?

When we look for words with a certain value, say 23, our thinking process might very well go through three stages.

Stage 1. We hunt for numbers that add up to 23. Here are some, written in, so-called, "tuple" form.

$(1,1,1,1,1,1,1,1,1,1,1,1,1,1,1,1,1,1,1,1,1,1,1,1)$
$(2,1,1,1,1,1,1,1,1,1,1,1,1,1,1,1,1,1,1,1,1,1,1,1)$
$(20,2,1)$
$(13,10)$
$(8,7,5,3)$

Stage 2. For each of these tuples, we match the numbers to the corresponding letters.

For the examples above, the tuples of corresponding letters are

(A,A,A,A,A,A,A,A,A,A,A,A,A,A,A,A,A,A,A,A,A,A,A,A)
(B,A,A,A,A,A,A,A,A,A,A,A,A,A,A,A,A,A,A,A,A,A,A,A)
(T,B,A)
(M,J)
(H,G,E,C)

Stage 3. Finally, for each tuple, we rearrange the letters and search for words.

In the examples above, we quickly discard all the tuples except (T,B,A) and (H,G,E,C) because it is very clear that the letters in the other tuples do not form words. With a little more thought and, perhaps, a dictionary, we can reject (H,G,E,C) and rearrange (T,B,A) to make BAT and TAB.

Let's make this into a game. You are encouraged to use a dictionary.

Part I: The Game of Wordsworth (Sum Version)

The One Player Game. Here are three ways to play.

1. Pick a number and challenge yourself to find a word with that value.

2. Pick a number and give yourself the challenge of finding as many words as you can with that value in, say, 5 or 10 minutes.

3. Pick a number and challenge yourself to find the longest possible word with that value. If you find a three letter word, look for a four letter word; if you find a four letter word, look for a five letter word and so on. It is very difficult to verify that you have found the longest word with that value. Someone with a larger dictionary might be able to find a longer one!

Now pick three numbers and try each version of the game. When you play, you will develop shortcuts and strategies.

The Group-Play Game. Here are two ways to play.

1. The number of rounds is equal to the number of players. The players take turns choosing a number to start a round. The first player to find a word whose value is equal to the number picked earns one point. The winner is the player with the most points after all rounds are complete.

2. The players agree on a number to start the game and set a time limit, such as 10 minutes. The winners are the players who find the most words with value equal to that number, or the longest word whose value is equal to that number.

This time, try playing the game with some friends. The more players there are, the more interesting the game because frequently everyone learns at least one new word.

Part II: The Game of Wordsworth (Product Version)

Now let's play the same game, but let's make it quite a bit more challenging. We are going to switch operations from addition to multiplication. This means that each letter of the alphabet has the same value as before, but each word has value equal to the product of the values of its letters. The game moves into the realm of much larger numbers very quickly.

For example, now "apples" has value $1 \times 16 \times 16 \times 12 \times 5 \times 19 = 291,840$ and Michael Jordan's name is now worth \$25,474,176,000 (which seems more appropriate than \$113).

As before, for the game, we first pick a number and then look for words whose value is that number. So, for example, if the number picked is 100, some words worth 100 are BABY which has value $2 \times 1 \times 2 \times 25 = 100$ and TEA which has value $20 \times 5 \times 1 = 100$.

Question 2. Can you find some other words worth 100?

Did you notice that in your hunt for words with a certain value you can put in as many of the letter A as you want?

Question 3. What is the special role played by the letter A?

The rules for this game are exactly the same as before, except that now the *value* of a word refers to the **product** of its letter values—not the **sum**.

The One Player Game. Here are three ways to play.

1. Pick a number and find a word with that value.

2. Pick a number and find as many words as you can with that value in, say, 5 or 10 minutes.

3. Pick a number and find the longest possible word with that value. Then find the shortest possible word with that value.

The Group-Play Game. Here are two ways to play.

1. The number of rounds is equal to the number of players. The players take turns choosing a number to start a round. The first player to find a word whose value is equal to the number picked earns one point. The winner is the player with the most points after all rounds are complete.

2. The players agree on a number to start the game and set a time limit such as 10 minutes. The winners are the players who find the most words with value equal to that number or the longest word whose value is equal to that number.

Play a few rounds of the game.

Are you ready for a challenge?

Wordsworth Challenge. Try to find one word for each of the numbers 10, 20, 30, 40, 50, 60, 70, 80, 90.

Now that you have become an adept Wordsworth player, here are some good questions to think about.

Question 4. Suppose you start with the number 17. Can you find any words whose value, using the product version of the game, is 17? What goes wrong?

Question 5. Suppose the number picked is 21 or 231. What goes wrong?

This game is more challenging than the sum version, don't you think? Often it is quite hard to find even one word, and, as you have seen, sometimes it is impossible to find any words at all. So let's take a good look at what is going on here.

When you look for words with a certain value, say 100, your thinking process may go through three stages.

Stage 1. The first task is to look for numbers no bigger than 26 whose product is 100. This means looking for *divisors*, or *factors*, of 100 which are no bigger than 26. When we write a number such as 100 as a product of some factors, we call the display a *factorization* of the number.

So, for example, here are all the factorizations of 100 with factors greater than 1 but no greater than 26:

$$100 = 5 \times 5 \times 4$$
$$100 = 5 \times 5 \times 2 \times 2$$
$$100 = 10 \times 5 \times 2$$
$$100 = 10 \times 10$$
$$100 = 20 \times 5$$
$$100 = 25 \times 4$$
$$100 = 25 \times 2 \times 2$$

What about 1? 1 is special. Since 1 is a factor of every number and since multiplying a number by 1 does not change the number, you can insert 1 anywhere you want and as often as you want in these factorizations.

Stage 2. Just as in the sum version of the game, you make the correspondence with letters:

(E,E,D,A,A,A,A,A,A,A, . . .)
(E,E,B,B,A,A,A,A,A,A, . . .)
(J,E,B,A,A,A,A,A,A,A, . . .)
(J,J,A,A,A,A,A,A,A,A, . . .)
(T,E,A,A,A,A,A,A,A,A, . . .)
(Y,D,A,A,A,A,A,A,A,A, . . .)
(Y,B,B,A,A,A,A,A,A,A, . . .)

Stage 3. For each tuple, you search for words using the letters in the tuple.

Here are some questions that will help you set up some strategies for the product version of the game.

Question 6. You probably know that a prime number is an integer, greater than 1, that has only 1 and itself as positive factors. Some examples of prime numbers are $3, 7, 17, 23, 47$, and 997. Which letters have prime number values?

Question 7. What are some numbers for which there are no words?

Question 8. Can you find any words whose value is a prime number?

Solutions

Question 1. Here are two more words with value 23: KEG and MADE.

Question 2. Two more 100 value words are DAY and EAT.

Question 3. A is special because it has value 1. The number 1 plays an exceptional role in multiplication because multiplying a number by 1 does not change the number. This means that A is a "free" letter in the game. You can use as many of the letter A as you want or need without changing the value of the product.

Wordsworth Challenge. One solution to the Wordsworth challenge is: BE with value 10; EBB with value 20; BOA with value 30; BAT with value 40; BAY with value 50; DO with value 60; BEG with value 70; DEAD with value 80; and CAFÉ with value 90.

Question 4. Suppose you start with the number 17. What goes wrong? 17 is a prime number. There is no way to write 17 as a product of numbers other than 17 itself and the number 1. A word with value 17 would have to consist of the letter Q and as many of the letter A as wanted. But there are no common English words of this form.

Question 5. Suppose the number picked is 21 or 231. What goes wrong? The words with value 21 would have to be made up from the letter U and as many of the letter A as wanted, or the letters G, C, and as many of the letter A as wanted. But, there are no common English words of this form. Words with value $231 = 3 \times 7 \times 11$ would have to be made up from the letters C, G, K and as many of the letter A as wanted, or U, K, and as many of the letter A as wanted. However, there are no common English words of this form.

Question 6. The letters B, C, E, G, K, M, Q, S, and W have prime number value.

Question 7. There are lots of numbers that are not the value of any common English word, for example, all of the (infinitely many!) prime numbers greater than 26.

Question 8. Here are four examples of words whose value is a prime number: BAA, MA, AM, and AS.

Suggestions for the Endurance Athlete

Use the product version of Wordsworth in the following challenges.

10K Challenge. Make it a class project to find words worth $30, 30^2, 30^3, 30^4, \ldots$. How high up can you go?

20K Challenge. Suppose n is a positive integer. Then n *factorial,* written $n!$, is the positive integer defined as follows:

$$n! = 1 \times 2 \times 3 \times (n - 1) \times n.$$

For example, $1! = 1$, $2! = 1 \times 2 = 2$, and $5! = 1 \times 2 \times 3 \times 4 \times 5 = 120$. Can you find a word with value $n!$ for each n up to $n = 5$? How about a word with value $n!$ for each n up to $n = 10$?

Bike

Numbers and Symmetry

Chapter Four

The Triangle and Square Games

These games are more elementary than magic squares. Similar skills are used, and, as for magic squares, players are led to consider whether a mathematical problem may have more than one solution, when two solutions are the "same," and whether all "different" solutions may be found.

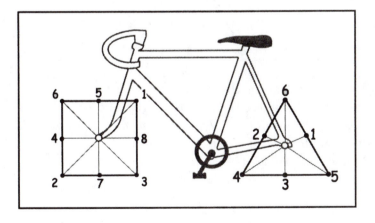

"There is nothing more productive of problems than a really good solution."

N. S. Kline

These games are fun and easy. All you need to begin playing is ordinary addition. However, you will learn how useful geometry can be to speed you along to the finish line.

Part I: The Triangle Game

To start, draw a triangle with all three sides equal (an equilateral triangle) and place a dot at each vertex and at the midpoint of each of the three sides.

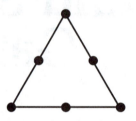

Now take the numbers 1, 2, 3, 4, 5, and 6 and place one number at each of the dots. If you do this haphazardly, you might get something like

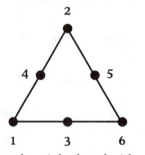

Observe that the sum along the right-hand side is 13, the sum along the left-hand side is 7, and along the bottom is 10. The object of The Triangle Game is to place the numbers $1, 2, 3, 4, 5,$ and 6 at the dots in such a way that the sums on all three sides are the same.

For example, the triangle

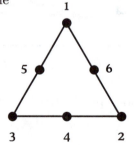

has the sum along each side, or *side sum*, equal to 9. A *solution* to the game is any triangle with side sums all equal. The triangle pictured is a solution with side sum equal to 9.

Question 1. Do you think there is a solution to the game with side sum equal to 1, 2, or 3? Why?

Directions for The Triangle Game

The Triangle Game has three stages.

Stage 1. Find a solution to the game (*other than the one given here earlier*).

To find a solution to the game means that you should place the numbers from 1 to 6 at the dots on the sides and at the vertices of your triangle so that the sum along each side is the same. Remember that you can use each number only once. Take some time now to try it.

Question 2. What side sum does your triangle have?

Stage 2. Find all numbers that can be side sums of solutions to the game.

One way of attacking this part is to argue, as you did earlier for 1, 2, and 3, that certain numbers are not possible as side sums. First look for the range of numbers that are potential side sums, and then verify whether they are actual side sums by finding solutions. If you would like some hints about potential side sums, keep on reading. If not, see if you can find the numbers on your own.

The hints come in the form of questions that will help you find the range of potential side sums.

Question 3. You know 9 is a possible side sum. Is 8 possible? Remember that the number 6 has to go somewhere. What about numbers smaller than 8?

Question 4. What about 13? Is it possible for 13 to be a side sum of a solution? Remember that 1 has to go somewhere. How about numbers larger than 13?

Whether you worked out the correct answers to Questions 3 and 4 on your own or consulted the Solutions, you now know that the possible side sums are the numbers 9, 10, 11, and 12. The next step is to try to find solutions with these numbers as side sums. You know there is a solution with side sum 9.

Question 5. Now try to find a solution for each of the side sums 10, 11, and 12. Give yourself plenty of time to work on this problem.

Did you find solutions for all four numbers 9, 10, 11, and 12? Excellent. There are many solutions for each side sum. This motivates Stage 3 of The Triangle Game. In Stage 3, geometry will give you the power and facility to find all solutions for each side sum in a very interesting way. You will also discover the relationship between solutions with the same side sum. In the meantime, one solution for each side sum can be found in the Solutions. Compare your solutions with the ones found there.

Solutions

Question 1. Is there a solution to the game with side sum equal to 1, 2, or 3? The answer is "No." Repeats aren't allowed, so it's impossible for the sum of three of the numbers 1, 2, 3, 4, 5, 6 to be 1, 2, or 3. You probably see that other numbers are not possible side sums. You will be able to use that information in The Triangle Game.

Question 2. It is one of the numbers 9, 10, 11, or 12.

Question 3. You know that 6 has to go on some side of the triangle with two other numbers between 1 and 5. The smallest numbers that can go with 6 are 1 and 2, so the smallest sum possible on that side is 9. This means that no number smaller than 9 can be a side sum for a solution.

Question 4. You know that 1 has to go on some side of the triangle with two other numbers between 2 and 6. The largest numbers that can go with 1 are 5 and 6, so the largest sum possible on that side is 12. Consequently, no number larger than 12 can be a side sum for a solution.

Question 5. Here is a set of solutions, one for each side sum. There are many other solutions.

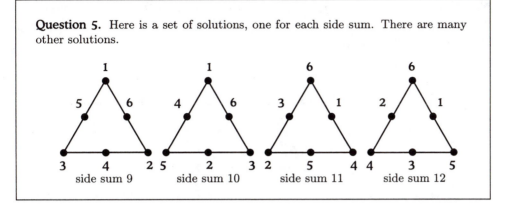

side sum 9   side sum 10   side sum 11   side sum 12

Stage 3. Find *all* solutions to the game.

You could search for all of the solutions by trial and error, but that wouldn't be very exciting. Also, how will you know when you have obtained all of them? Instead, let's introduce some geometry. Geometry will provide a very efficient technique for morphing one solution into another with the same side sum.

Using geometry for Stage 3. Take the triangle T below with side sum 9 and rotate the triangle 120° clockwise so that the vertex where the 1 is goes to the vertex where the 2 is, like this:

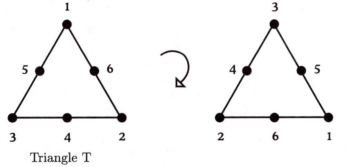

Triangle T

The side sum of this new triangle is still 9 but the numbers are in different places. So we have a new solution. Notice that this solution is similar to the old one because when you read the numbers, starting with 1, and move clockwise around the triangle, the order is the same, namely 1, 6, 2, 4, 3, 5.

Question 6. There is a different way to rotate the triangle to get an additional solution with side sum 9. Can you find it?

Next comes another geometric idea to find still other solutions. Take triangle T with side sum 9 and draw a line from the top vertex to the midpoint of the bottom side, that is from 1 to 4.

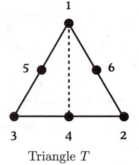

Triangle T

Flip the triangle around this line to get:

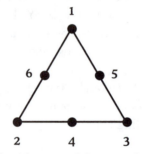

This is a new solution with side sum 9 but notice that to get the numbers in the same order as before (1, 6, 2, 4, 3, 5), you must move around the triangle in the counterclockwise direction.

Question 7. There are two more flips, or reflections, that result in two more solutions with side sum 9. Can you find them?

These six solutions (triangle T, the two rotated triangles, and the three flipped triangles) are all the solutions with side sum 9. It is not difficult to verify this. If you are interested, you can work out the ideas in the Heavy Lifting section later.

Observe the power geometry has added to the game. One solution immediately gives five others, and these six are all the solutions with side sum 9.

You can apply this power and efficiency to find all solutions for other side sums. You are on your own now to finish the game.

Did you discover that there are exactly twenty-four solutions to The Triangle Game? If so, you have completely solved the game. That is quite an accomplishment.

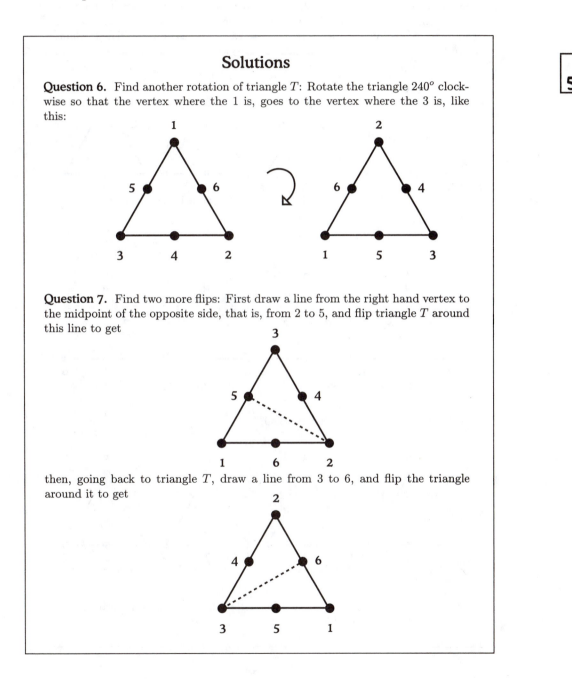

Solutions

Question 6. Find another rotation of triangle T: Rotate the triangle $240°$ clockwise so that the vertex where the 1 is, goes to the vertex where the 3 is, like this:

Question 7. Find two more flips: First draw a line from the right hand vertex to the midpoint of the opposite side, that is, from 2 to 5, and flip triangle T around this line to get

then, going back to triangle T, draw a line from 3 to 6, and flip the triangle around it to get

For each possible side sum 9, 10, 11, and 12, once you have one solution, there will be two more solutions given by rotations (one through 120° and one through 240°) and there will be three more solutions given by flips (one around each of the *medians* of the triangle, or the lines that connect a vertex with the midpoint of the opposite side). Thus, for each of the four possible side sums, there are six solutions. So, all in all, there are $4 \times 6 = 24$ solutions to The Triangle Game. Here they are:

side sum 9   side sum 10   side sum 11   side sum 12

Rotate the triangles in the first row 120°:

side sum 9   side sum 10   side sum 11   side sum 12

Rotate the triangles in the first row 240°:

side sum 9   side sum 10   side sum 11   side sum 12

Flip the triangles in the first row around the line from the top vertex to the midpoint of the opposite side:

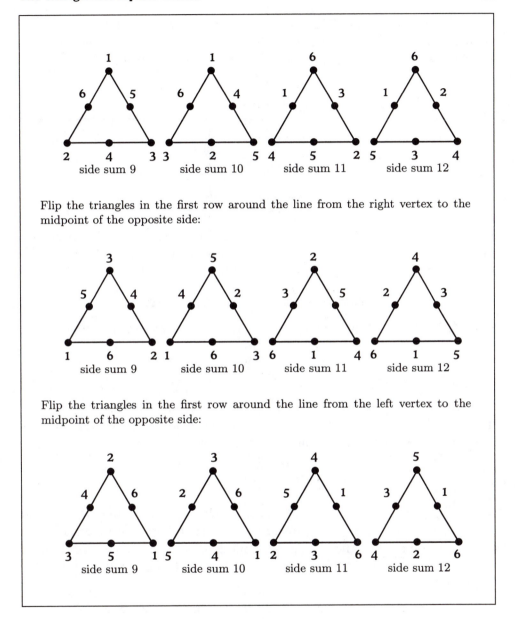

Flip the triangles in the first row around the line from the right vertex to the midpoint of the opposite side:

Flip the triangles in the first row around the line from the left vertex to the midpoint of the opposite side:

Part II: The Square Game

What's next? Squares, of course. By now, you can probably set up the whole game and analyze it yourself. Be on the lookout for some surprises.

For a square, there are eight dots and the numbers from 1 to 8 are used. (Some of you may recognize the similarity to Magic Squares except that in The Square Game there is no dot in the middle of the square.)

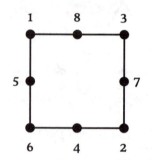

Just as for The Triangle Game, a solution to The Square Game is any square with side sums all equal. The square pictured is a solution with side sum equal to 12.

Directions for The Square Game

Stage 1. Find a solution for The Square Game (other than the one given).

Stage 2. Find all possible side sums. (Watch for some surprises.)
If you would like some hints for this part, keep reading. Otherwise, set out on your own but be alert for hazards on the road ahead.

Backed by your experience and training in The Triangle Game, it ought to be easy to decide which side sums are not possible for squares. First of all, 8 must go somewhere and the smallest numbers that can share a side with 8 are 1 and 2 so, for a square, you can't have a side sum of 10 or less. But (surprise!):

> A solution with side sum 11 is not possible.

Can you figure out why? It takes a few steps to explain this so let's work it out together. The 8 has to go somewhere. It can go either at a vertex or at a midpoint. Try a vertex first. You have to use 1 and 2 along with 8 to get a side sum of 11.

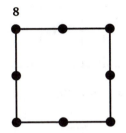

Say 1 and 2 go on the top with 8. Then what numbers go along the left side?

There are no numbers remaining to go along the left side. This means that 8 cannot be at a vertex. Now suppose 8 is at the midpoint of the top. Again, 1 and 2 have to go with 8. So 1 and 2 are at the top vertices, with, say, 1 on the left and 2 on the right.

Question 8. What can go with 1 along the left side?

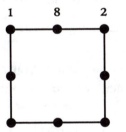

Question 9. What can go with 2 along the right side?

Question 10. If you put 7 and 3 on the left, what has to go on the right?

Question 11. Now for the clincher. What is the only number remaining to go along the bottom? What goes wrong?

So, as you can see from the Solutions, 7 and 3 cannot go on the left.

Question 12. Suppose instead that you put 6 and 4 on the left. What can go on the right?

Nothing is available to go on the right. Consequently, there can be no solution with side sum 11, and the smallest potential side sum is 12. However, 12 is an actual side sum, as our first Square Game example shows.

How about the biggest possible side sum? Since 1 has to go somewhere and the largest numbers that can share a side with 1 are 7 and 8, it follows that 16 is the largest potential sum. But, just as above:

A solution with side sum 16 is not possible.

Question 13. Can you figure out why? You are on your own for this one. Try the same type of argument we worked out together above.

This leaves possible side sums 12, 13, 14, and 15.

Question 14. Can you find solution squares with side sums of 12, 13, 14, and 15?

Stage 3. Take one of your solutions to The Square Game and find the number of solutions you get by applying rotations and flips to it.

For example, look at the solution

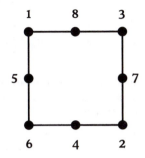

with side sum 12. Rotating the square clockwise through 90°, 180° and 270° gives three additional solutions.

Question 15. There are also four flips that give four more solutions. What are they?

Did you notice that you weren't asked, as you were in The Triangle Game, to find all solutions to The Square Game? For The Triangle Game, it is true that all solutions are obtained by taking one solution for each side sum and applying rotations and flips. This is no longer true for The Square Game. (Mathematics is full of surprises!) If you are interested in finding all solutions for The Square Game, see the Heavy Lifting section.

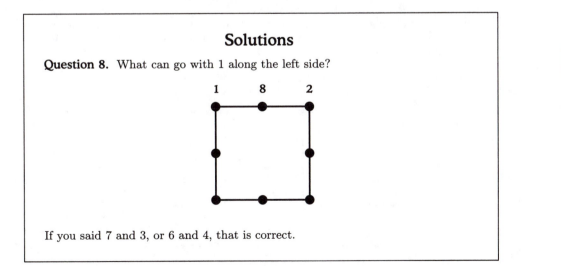

Solutions

Question 8. What can go with 1 along the left side?

If you said 7 and 3, or 6 and 4, that is correct.

Question 9. What can go with 2 along the right side? Did you say 6 and 3, or 5 and 4? Good.

Question 10. Now if you put 7 and 3 on the left, what has to go on the right? The answer is 5 and 4.

Question 11. What is the only number remaining to go along the bottom? What goes wrong? The number 6 remains. To get a sum of 11 you would need 4 and 1, or 3 and 2, but neither of those pairs is available.

Question 12. Suppose instead that you put 6 and 4 on the left. What can go on the right? Nothing! Neither of the pairs 6 and 3, or 5 and 4 is available. This completes the proof that the smallest potential side sum is 12.

Question 13. Show that there is no solution with side sum 16: There is no way to complete the square if 1 and 7 and 8, in any order, are on one side, say the top side. 1 can't go at a vertex because you can only use the 7, 8 pair once. If 1 goes at the midpoint, then 7 is at one vertex, say the left, and 8 is at the other, the right.

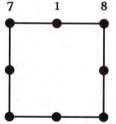

The possible pairs with 7 are 6 and 3, or 5 and 4. The possible pairs with 8 are 6 and 2, or 5 and 3. But you can't put the pair 6, 3 on the left because then no pair is available for the right. If the pair 5, 4 goes on the left, then the pair 6, 2 must go on the right. That leaves 3 for the bottom. Now for the bottom side sum to be 16, either the pair 8, 5 or the pair 7, 6 must go with 3, but neither pair is available.

Question 14 Here are some solutions to The Square Game.

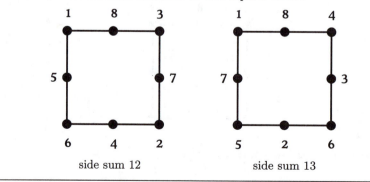

side sum 12   side sum 13

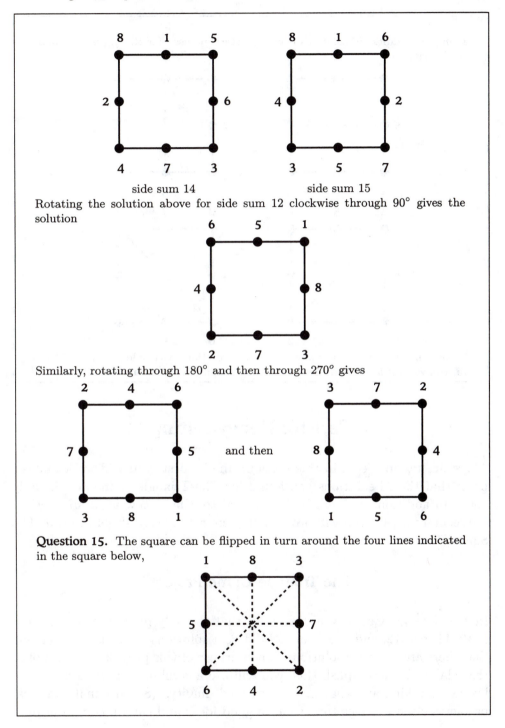

side sum 14   side sum 15

Rotating the solution above for side sum 12 clockwise through 90° gives the solution

Similarly, rotating through 180° and then through 270° gives

and then

Question 15. The square can be flipped in turn around the four lines indicated in the square below, giving four additional solutions from flips (the line used for the flip is indicated on the square):

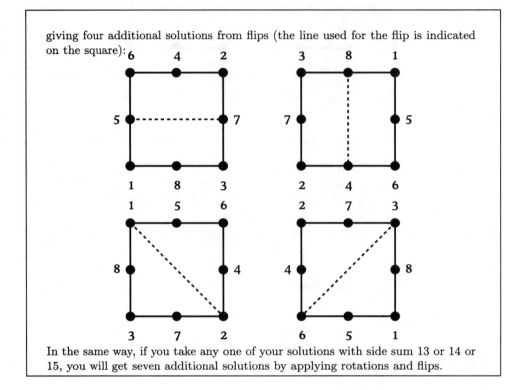

In the same way, if you take any one of your solutions with side sum 13 or 14 or 15, you will get seven additional solutions by applying rotations and flips.

Part III: Heavy Lifting

There are two projects in this section. In the first, you will work out the proof that the 24 solutions you found for The Triangle Game are the only ones. In addition, you will be introduced to a neat new idea: *duality*. In the second project, you will prove that there are exactly 48 solutions to The Square Game.

The Triangle Game Project

In The Triangle Game, you discovered that the only possible side sums are 9, 10, 11, and 12, and you found 24 different solutions. The claim was made that there are no other solutions. The purpose of this project is the proof of that claim. To accomplish this, you will use a combination of careful case-by-case checking and that new idea called duality. (Solving mathematical problems often involves this blend of good ideas and careful computation.)

Why are there exactly 24 solutions to
The Triangle Game?

The idea is to show that every solution, for a fixed side sum, is obtained from any one solution by a rotation or a flip (reflection). This means that you may employ rotations and reflections (flips) in your arguments anytime it is helpful.

First, the case work.

Side Sum 9. Tackle the case where the side sum is 9. The number 1 can go either at a vertex or a midpoint, so assume to start that 1 is at a vertex. What numbers can share a side with 1? Either 2 and 6, or 3 and 5. So, since rotations and reflections are allowed, you can assume that 1 is at the top vertex, and the pair 2 and 6 go along the right side, and the pair 3 and 5 go along the left side.

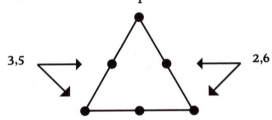

Notice that 4 is the only number remaining to go at the midpoint of the bottom.

Question 16(a). Can 6 be at the vertex on the right?

Question 16(b). Can 5 be at the vertex on the left?

Question 17. Once you establish that the answer to both of the questions above is "no," you must show that there is no side sum 9 solution with 1 at a midpoint.

The argument that 1 cannot be at a midpoint for a side sum 9 solution completes the proof that the only side sum 9 solutions are the same six known solutions obtained by rotation and reflection.

Side Sum 10. Here is a solution with side sum 10.

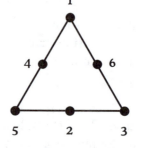

Question 18(a). For the case of side sum 10, you are on your own to show, just as you did above, that the only solutions with 1 at a vertex and side sum 10 are this one and the five others obtained from it by rotations and flips.

Question 18(b). To complete the side sum 10 case, show that 1 cannot be at a midpoint.

Here is a neat payoff for all your hard work. You can use a method called *duality* to handle the cases for side sums of 11 and 12. You are going to show that the following statements are true.

> Every side sum 9 solution triangle corresponds to a side sum 12 solution triangle and vice versa.

> Every side sum 10 solution triangle corresponds to a side sum 11 solution triangle and vice versa.

To begin, notice that the sum of all the numbers around the triangle is 21, i.e., $1+2+3+4+5+6 = 21$. Now suppose you have a solution triangle T with a certain side sum s, and the numbers along one side are a, b, and c, so $s = a + b + c$. Take another triangle, and, on the same side, replace a by $7 - a$, b by $7 - b$ and c by $7 - c$.

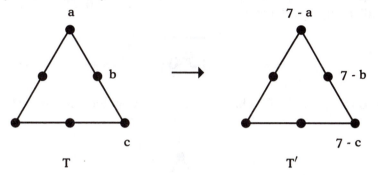

What is the sum along this side of the new triangle T'?

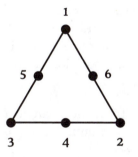

Did you see that the side sum is $(7-a)+(7-b)+(7-c) = 21 - (a+b+c) = 21 - s$? Good.

How does this work out for the triangle below?

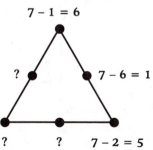

If you apply this process you will get a new triangle, called the *dual triangle*. It looks like this:

Question 19. Try to continue the process of substraction from 7 for the remaining dots on the triangle. Is the dual triangle a solution to the game? What side sum does it have?

Return to the triangle T, where all the dots now have letters attached.

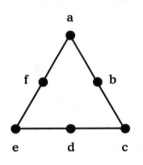

Get ready for some interesting challenges. Since this is heavier lifting, we'll introduce some mathematical terms related to the word "dual." The challenges will establish a correspondence, called *duality*, between solution triangles of side sum 9 and solution triangles of side sum 12, and between solution triangles of side sum 10 and solution triangles of side sum 11. Notice that $12 = 21 - 9$ and $11 = 21 - 10$.

The process of going from the triangle T with numbers a, b, c, d, e, f to the dual triangle T' with the numbers $7 - a, 7 - b, 7 - c, 7 - d, 7 - e, 7 - f$ is called the *dualizing process*.

Challenge 1. Assume the triangle T is a solution to the game with side sum s. Show that the dual triangle T' is a solution with side sum $21 - s$.

Can you guess what will happen if you repeat the dualizing process? In other words, take each number, say $7 - a$, for example, on the dual triangle T', subtract it from 7 and form a new *double-dual* triangle.

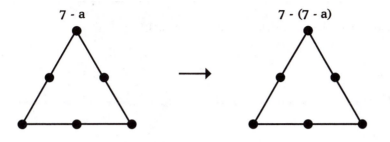

7 - a   7 - (7 - a)

Challenge 2. Repeat the dualizing process on the dual triangle T'. What triangle do you get?

Challenge 3. Can you conclude that all solutions with side sum 11 and 12 are obtained from solution triangles with side sum 9 and 10? Explain why.

If you answered the three challenges, you have succeeded at some very heavy lifting. That is quite an accomplishment!

Solutions

Question 16. The answer to both questions is "no" because you cannot get a side sum of 9 with either 6 or 5 at a vertex and 4 at the midpoint. So if 1 is at a vertex, the only solutions you can get are

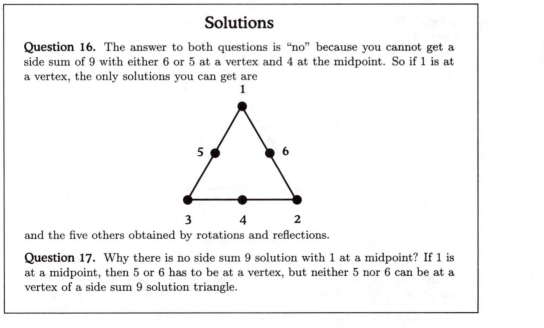

and the five others obtained by rotations and reflections.

Question 17. Why there is no side sum 9 solution with 1 at a midpoint? If 1 is at a midpoint, then 5 or 6 has to be at a vertex, but neither 5 nor 6 can be at a vertex of a side sum 9 solution triangle.

Question 18. Here is a solution with side sum 10.

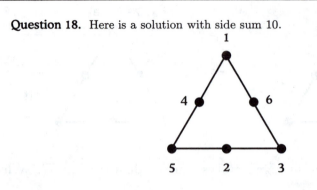

(a). Why are this solution and the five others obtained from it by rotations and reflections the only solutions with side sum 10? The reasoning follows the same pattern as for the case of side sum 9. Assume first that you have a solution with 1 at vertex. Then the only pairs of numbers that can go along a side with 1 are 5, 4 and 6, 3. Since rotations and reflections are allowed, you may assume that 1 is at the top vertex, 5 and 4 are on the left side, and 6 and 3 are on the right. The only number remaining to go at the midpont of the bottom is 2. But with 2 at the midpoint of the bottom, the only pair of numbers that can go with 2 are 5 and 3. This means that neither 4 nor 6 can be at a vertex. So 4 has to go at the midpoint of the left side, and 6 has to go at the midpoint of the right side. (b). Why can't 1 go at a midpoint? If it does, then 1 would have to be paired with either 5 and 4 or with 6 and 3. But then either 4 or 6 would be at a vertex. That cannot happen, because in the first case, 5 and 1 are not available, and, in the second case, 3 and 1 are not available.

Question 19. Is the dual triangle a solution to the game? What side sum does it have?
The dual triangle looks like this:

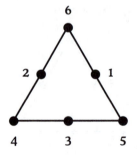

The dual triangle is a solution with side sum 12.

Challenge 1. Assume the triangle T is a solution to the game with side sum s. Show that the dual triangle T' is a solution with side sum $21 - s$.

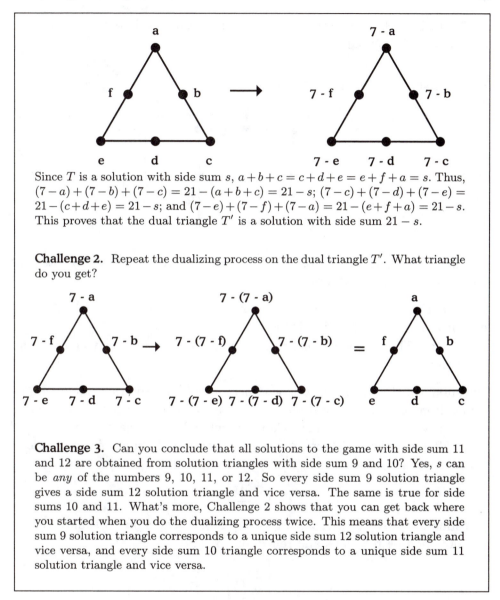

Since T is a solution with side sum s, $a+b+c = c+d+e = e+f+a = s$. Thus, $(7-a)+(7-b)+(7-c) = 21-(a+b+c) = 21-s$; $(7-c)+(7-d)+(7-e) = 21-(c+d+e) = 21-s$; and $(7-e)+(7-f)+(7-a) = 21-(e+f+a) = 21-s$. This proves that the dual triangle T' is a solution with side sum $21-s$.

Challenge 2. Repeat the dualizing process on the dual triangle T'. What triangle do you get?

Challenge 3. Can you conclude that all solutions to the game with side sum 11 and 12 are obtained from solution triangles with side sum 9 and 10? Yes, s can be *any* of the numbers 9, 10, 11, or 12. So every side sum 9 solution triangle gives a side sum 12 solution triangle and vice versa. The same is true for side sums 10 and 11. What's more, Challenge 2 shows that you can get back where you started when you do the dualizing process twice. This means that every side sum 9 solution triangle corresponds to a unique side sum 12 solution triangle and vice versa, and every side sum 10 triangle corresponds to a unique side sum 11 solution triangle and vice versa.

The Square Game Project

The fact that every solution to The Triangle Game is obtained by rotation or reflection from a basic set of four triangles, one for each side sum, is no longer true for The Square Game. Not all solution squares are obtained by rotation and flips from a set of four solutions, one for each side sum of 12,

13, 14, and 15. This investigation will lead to new solutions and to the total number of solution squares.

Here is a solution square S with side sum 13:

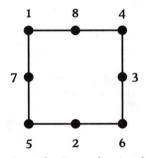

You know there are seven more solutions obtained from this one by rotations and reflections for a total of eight solution squares.

Stage 1. Find a solution to The Square Game for side sum 13 that is not obtained by rotations and reflections of S.

Hint: Think about the following questions before you look for an example.

Question 20. How can you tell whether or not one solution square is obtained from another by rotations or reflections? The answer to the following question will give you a clue. Can a number be at a vertex to start and then, after rotations and/or reflections, end up at the midpoint of a side?

Question 21. Next, search for an example of a side sum 13 square that is not obtained by rotations and reflections of S.

Question 22. If you apply rotations and reflections to the new square you found, how many new side sum 13 solution squares do you obtain?

At this point, you should have a total of sixteen squares with side sum 13.

Now you have some work to do to account for *all* of the solutions with side sum 12 and 13.

Challenge for Side Sum 12. Show that the side sum 12 solution square,

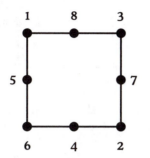

and the seven other solution squares obtained by rotations and reflections are the only solution squares of side sum 12.

Hint: Use the technique of analyzing possible pairs on a side that you used earlier to show there are no solution squares of side sum 11 or 16.

Challenge for Side Sum 13. Show that the 16 solution squares you found are the only solution squares with side sum 13.

Hint: The same hint applies.

Can you predict what will happen for solution squares of side sum 14 and 15? To find out if you are correct, let's see what duality for squares looks like.

Stage 2. Duality for Squares.

For squares, the sum of all the integers around the square is 36, so since the square has four sides, we will use $\frac{36}{4} = 9$ instead of the $7 = \frac{21}{3}$ that we used for triangles.

Start with the square:

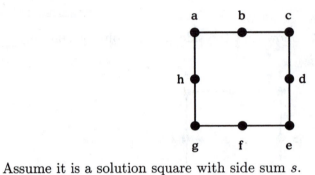

Assume it is a solution square with side sum *s*.

Form the dual square:

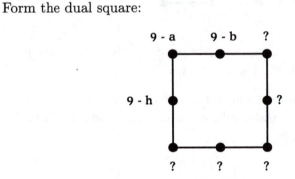

Here are the crucial duality questions.

Duality Question 1. Is the dual square a solution? If so, what is its side sum?

Duality Question 2. Repeat the dualizing process on the square. What square do you get?

Duality Question 3. Can you conclude that all solutions to the game with side sum 14 and 15 are obtained from solution squares with side sum 13 and 12? Explain why.

Final Challenge. Put Parts 1 and 2 of the project together and use duality to show that there are exactly 48 solutions to The Square Game.

Solutions

Question 20. A number can not be at a vertex to start, and then, after rotations and/or reflections, end up at the midpoint of a side? To justify this answer, check all three rotations and all four reflections to see that a number at a vertex ends up at a vertex and a number at a midpoint ends up at a midpoint.

Question 21. Here is one solution for an example of a side sum 13 that is not obtained by rotations and reflections of S.

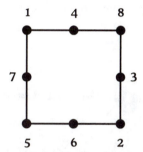

It has 8 at a vertex so it cannot be obtained by rotations and flips from the original side sum 13 square which has 8 at a midpoint.

Question 22. If you apply rotations and reflections to the square, you found, as above, you will obtain seven new ones.

Challenge for Side Sum 12. The arguments are similar to the ones used earlier to show that solutions with side sum 11 and 16 are impossible. Suppose 1 is at a vertex. The only number pairs that can go with 1 are $8, 3$; $7, 4$; and $6, 5$. Since rotations and flips are allowed, you can assume 1 is at the top left vertex. Note that the pair $3, 1$ is the only one that can go with 8 to get side sum 12. This means that 8 cannot be at a vertex and 1 must be paired with 8 and 3. Since reflections are allowed, you can assume that $1, 8, 3$ are on the top in that order.

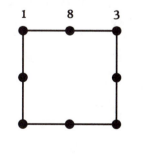

Suppose the pair 7, 4 is on the left. Then no pairs are available to go with 3 to give side sum 12 on the right. So the pair 6, 5 must be on the left, and the only pair remaining to go with 3 on the right is 7, 2. This puts the number 4 at the midpoint of the bottom. Then to get side sum 12 along the bottom, you see that 6 must be at the bottom left vertex and 2 at the bottom right. This means that the only side sum 12 solutions with 1 at a vertex are the ones named in the problem.

Now check that the number 1 cannot be at a midpoint. The number 8 has to go somewhere. Remember that 3, 1 is the only pair that can go with 8. This means that 8 cannot be at a vertex. But if 1 is at a midpoint, then 8 has to be at a vertex.

Challenge for Side Sum 13. The arguments are the same.

Square Duality Questions and the Final Challenge. These solutions are analogous to those for triangle duality.

Suggestions for the Endurance Athlete

Each one of the following projects will produce surprises.

10K Challenge. Investigate the pentagon game.

20K Challenge. Investigate the hexagon game.

30K Challenge. Investigate the n-gon game. Start this one by finding one solution for each n.

Chapter Five

Palindromes

This investigation prevents aibohphobia—the fear of palindromes.

"I prefer pi."

Anonymous

What is a palindrome? The usual meaning of palindrome is a word or phrase that reads the same foward or backward. Familiar examples are: dad, toot, radar, and 'Madam I'm Adam.' The word "palindrome" comes from the Greek word "palindromos" and means "running back again." Since numbers can also "run back again," we will use the word palindrome to mean a number whose digits read the same forward or backward. All numbers will be written in base 10.

Examples of numbers that are palindromes are easy to find. All one-digit numbers are palindromes. That gives us ten examples (including the number 0) of palindromes. Here are some other examples: $33, 141, 5775,$ and 135797531.

Let's think about dates that are palindromes. Write each date as a sequence of eight numbers $abcdefgh$, where a and b are the digits of the month, c and d are the digits of the day, and $e, f, g,$ and h are the digits of the year. For example, write January 5, 1956 as 01051956, and November 3, 2001 as 11032001. Notice that October 2, 2001 corresponds to the palindrome 10022001. Can you find some other dates that are palindromes? What are some examples of years for which none of the dates is a palindrome?

Part I: Palindromes of Five or Fewer Digits

Number palindromes are very interesting. It is always a good idea to begin an inquiry with lots of examples. Your palindrome investigation will begin with the hunt for all palindromes with five or fewer digits. I'll help you get started by asking some questions and giving some hints. Try to answer each question in turn because each answer depends on the ones that come before. If you have thought long and hard about a question and are still stuck, you can look at the solutions for help.

Two-Digit Palindromes

Question 1. Can you find all the two-digit palindromes? How many are there?

Three-Digit Palindromes

A three-digit palindrome has to have the same number as its first and last digit but can have any number between 0 and 9 as its middle digit. So, starting with the two-digit palindrome 55, for example, you can form the three-digit palindrome 505. See if you can use this idea to answer the following question.

Question 2. Start with any one of the two-digit palindromes you found. How many three-digit palindromes can you make from it? Here is a partially completed table to help you get started.

Two-Digit Palindromes	Three-Digit Palindromes
11	$101, 111, 121, \ldots$
22	$202, 212, \ldots$
33	

Using two-digit palindromes to make three-digit palindromes.

You will need to use your answer to Question 1 to answer the next question.

Question 3. How many three-digit palindromes do you get from all of the two-digit palindromes?

Question 4. Are there any other three-digit palindromes?

Four-Digit Palindromes

Question 5. How many four-digit palindromes can you construct from the 9 two-digit palindromes? Here is a table to help you get started.

Two-Digit Palindromes	Four-Digit Palindromes
11	$1001, 1111, 1221, 1331, \ldots$
22	$2002, 2112, \ldots$
33	

Using two-digit palindromes to make four-digit palindromes.

Question 6. Are there any other four-digit palindromes?

Now you are into the swing of this. Can you see how to use these ideas to construct all the five-digit palindromes?

Five-Digit Palindromes

Question 7. How many five-digit palindromes are there?

Hint: What can you say about the outer two digits of a five-digit palindrome? What about the middle three digits of a five-digit palindrome?

Be careful here. Are the middle three digits of a five-digit palindrome always a three-digit palindrome?

Two-Digit Palindromes	Five-Digit Palindromes
11	$10001, 10101, 10201, \ldots, 11011, 11111, \ldots, \ldots, 12021, \ldots$
22	$20002, 20102, \ldots$
33	

Using two-digit palindromes to make five-digit palindromes.

Congratulations! You now know all numbers with five or fewer digits that have the property of being palindromes.

Solutions

Question 1. Did you find the nine two-digit palindromes: $11, 22, 33, 44, 55, 66, 77, 88, 99$?

Question 2. For each two-digit palindrome, such as 33, there are 10 three-digit palindromes, 303, 313, 323, 333, 343, 353, 363, 373, 383, 393, constructed by inserting the numbers 0, 1, 2, ..., 9 in turn between the original two digits.

Question 3. Remember that you counted nine two-digit palindromes. For each of them you can build 10 three-digit palindromes. So there are $9 \times 10 = 90$ three-digit palindromes constructed in this way.

Question 4. No. Since the numbers "running forward" and "running back again" have to be the same, the first digit and the last digit of every three-digit palindrome must be the same.

Question 5. For each two-digit palindrome, such as 88, there are 10 four-digit palindromes, 8008, 8118, 8228, 8338, 8448, 8558, 8668, 8778, 8888, 8998, constructed by inserting all the two-digit palindromes as well as the digits 00 in turn between the original two digits. This gives $9 \times 10 = 90$ new examples, all with four digits. So the answer to Question 5 is 90.

Question 6. No. A four-digit palindrome must have the same first and last digits and the same middle two digits.

Question 7. There are 900 five-digit palindromes. They can be constructed and counted this way. You know that a five-digit palindrome must have first and last digits the same, and second and fourth digits the same with any number between 0 and 9 as the middle digit. So the outer two digits of a five-digit palindrome form a two-digit palindrome. You have to be a little careful about the middle three digits. As long as the second and fourth digits are not equal to 0, the middle three digits form a three-digit palindrome. So you can construct most of the five-digit palindromes by taking each of the nine two-digit palindromes and inserting, in turn, all of the 90 three-digit palindromes in between. This gives $9 \times 90 = 810$ five-digit palindromes such as 11011, 24542, 53835, etc. To get the remaining five-digit palindromes, insert $000, 010, 020, 030, 040, 050, 060, 070, 080, 090$, in turn, between each of the nine two-digit palindromes. This gives $9 \times 10 = 90$ more five-digit palindromes. The sum total is then 900.

Suggestions for the Endurance Athlete

10K Challenge. Work with some of your classmates to find a formula for the number of d-digit palindromes for $d > 5$.

20K Challenge. You have figured out the number of d-digit numbers for $d \leq 5$. Do you see a pattern that might help you guess the number of d-digit palindromes for $d > 5$? You might want to try $d = 6$ and 7 before undertaking general d.

Part II: The Reverse and Add Rule

You saw that only nine two-digit numbers are palindromes. What about all of the other two-digit numbers that are not palidromes? In Part III, you will show that all two-digit numbers are just a few steps away from being palindromes. This involves a neat mathematical procedure called the *reverse and add rule*.

The number 25 is not a palindrome, but here is a mathematical procedure that turns it into one. Take the two digit number 25, *reverse* the digits to get the number 52 *and add* the two numbers together. What do you get?

You get the palindrome $77 = 25 + 52$. Is this just because 25 is a special number? Try the reverse and add rule on some other two-digit numbers and see. Try $13, 45$, and 56.

It works! You get $13 + 31 = 44, 45 + 54 = 99, 56 + 65 = 121$, all palindromes. Now try 67.

Unfortunately, $67 + 76 = 143$, and 143 is not a palindrome.

The saying "If at first you don't succeed, try, try again," may not be a palindrome, but it's a good motto for many endeavors, particularly, mathematics. Try the reverse and add rule again:

$$67 + 76 = 143;$$
$$143 + 341 = 484;$$

and 484 is a palindrome. Do you think you are on to something here?

Try the number 10.

The number 10, as well as each of the other two-digit numbers with units digit equal to 0, is special because the number obtained by reversing digits is not a two-digit number. That's O.K. As you can see, the reverse and add rule still works, and it works on the first try: $10 + 01 = 10 + 1 = 11$, a palindrome.

Question 8. Does it also work for all of the other two-digit numbers that have units digit equal to 0?

Question 9. Next try the number 78, and don't forget the motto "If at first you don't succeed, try, try again."

Do you think the reverse and add rule might eventually turn every two-digit number into a palindrome? The answer to this question will be found in Part III.

First, there is some preparation to do. Let's begin by giving a name to numbers that become palindromes using the reverse and add rule. Let k be a non-negative integer. A number N will be called a *k-step palindrome* if the reverse and add rule applied to N results in a palindrome after k applications or steps and no fewer. The original palindromes, numbers such as 66, are zero-step palindromes. You have shown that 25 is a one-step palindrome:

$$\text{first step: } 25 + 52 = 77.$$

You have also shown that 67 is a two-step palindrome:

$$\text{first step: } \quad 67 + 76 = 143;$$
$$\text{second step: } \quad 143 + 341 = 484.$$

What step palindrome is 78?

Here are two questions that will aid in the investigation of two-digit k-step palindromes.

Question 10. You know the number 78 is a four-step palindrome. What step palindrome is 87?

Think! Don't compute to answer this question.

Now be careful with the next question. Remember that the first digit of a non-zero number is not equal to zero.

Question 11. If a two-digit number is a k-step palindrome, is the number obtained by reversing its digits also a k-step palindrome?

As you can see from the Solutions, the answer to Question 11 depends on whether the units digit is zero or not zero.

Let's generalize Question 11. Suppose you have a d-digit number N and suppose that N is a k-step palindrome. Is the number N' obtained by reversing the digits of N also a k-step palindrome?

Just as for Question 11, the answer depends on whether the units digit of N is zero or not. If the units digit of N is 0, then the number N' obtained by reversing digits will have fewer digits and no conclusion about the number of steps can be drawn. For example, if $N = 270$, then N is a two-step palindrome. However, $N' = 027 = 27$ is a one-step palindrome. If the units digit of N is not 0, then the number N' has the same number of digits. In this case, if N is a palindrome (a zero-step palindrome), then, of course, so is N'. If N is not a palindrome, then N' is also not a palindrome. Both numbers have the same first step, and as each step depends only on the one immediately before it, each succeeding step will also be the same. So the answer to the question is "yes," in this case. This handy fact is a big time saver. You get "2 for 1."

> If a number N is a k-step palindrome and
> if the units digit of N is not equal to 0,
> then the number N'
> obtained by reversing the digits of N
> is also a k-step palindrome.

Ideas, like this one, which lead to methods for dealing with more than one number at a time, illustrate the power of mathematical thinking.

Solutions

Question 8. Yes, it does work for all of the other two-digit numbers with units digit equal to 0. In fact, you get all the two-digit palindromes this way: $20+02 = 22, 30+03 = 33, 40+04 = 44, 50+05 = 55, 60+06 = 66, 70+07 = 77, 80+08 = 88$ and $90 + 09 = 99$.

Question 9. Did you see that persistence pays off? After four applications of the reverse and add rule, the palindrome 4884 is achieved:

$$\text{Step 1:} \quad 78 + 87 = 165;$$
$$\text{Step 2:} \quad 165 + 561 = 726;$$
$$\text{Step 3:} \quad 726 + 627 = 1353;$$
$$\text{Step 4:} \quad 1353 + 3531 = 4884.$$

This makes 78 a four-step palindrome.

Question 10. You are right if you said four. The numbers 87 and 78 have the same first-step number, namely 165. Each succeeding step uses only the step before, so since 78 is a four-step palindrome, 87 is also.

Question 11. If the number obtained by reversing digits is a two-digit number, then the answer is "yes." For if the number is a palindrome, then, of course, so is the number obtained by reversing digits. If the number is not a palindrome, then neither is the number obtained by reversing digits. Both numbers will have the same first step, and each succeeding step will also be the same. Note that it is important to require that the number obtained by reversing digits has two digits. For example, 10 is a one-step palindrome but $01 = 1$ is a zero-step palindrome. Requiring that the number obtained by reversing digits has two digits means that the units digit of the original number is not 0.

Part III: The Two-Digit Palindrome Game

Are you ready for the two-digit palindrome game? The object of the game is to take each of the two-digit numbers and to figure out what step (if any) palindrome it is.

There's an advantage to thinking about what it means to write a two-digit number in base 10. Here is a little review of that.

Every two-digit number N has a tens digit t and a units digit u, where t is an integer between 1 and 9 and u is an integer between 0 and 9, and $N = 10t + u$. For example, $45 = 10(4) + 5$ has tens digit 4 and units digit 5, and $20 = 10(2) + 0$ has tens digit 2 and units digit 0. (Be careful. 45 can also be written: $45 = 10(3) + 15$, but you know 15 is not the units digit because it is greater than 9, so 3 is not the tens digit either.) Here's a good question to test your "digit dexterity."

Digit Dexterity Question. If $N = 10t + u$, what does the number N' obtained by reversing digits look like?

Directions for The Two-Digit Palindrome Game

Object of the Game. Take the collection of all 90 two-digit numbers. For each one, find out what step, if any, palindrome it is. Make a table of all zero-step palindromes, one-step palindromes, two-step palindromes, etc., as far out as you need to go so that all two-digit numbers appear.

Scoring. 0 points for each zero-step palindrome, 1 point for each one-step palindrome, and k points for each k-step palindrome.

Time Limit. 15 minutes for your first attempt.

Directions. On your first attempt, fill in as much of the table as you can in 15 minutes. Then work through the Post-Game Analysis below. The mathematical perspective gained there will give you a big boost towards completing the table the second time around. Your final score will be tallied after you have had two opportunities to work on the table.

An important suggestion before you begin. You can work on one number at a time. However, you are encouraged to apply your powers of analysis and deduction to tackle the question more broadly. Try to work with the base 10 representation of the numbers. For example, if you start with the two digit number $N = 10t + u$, can you figure out the number that you get at the first step in base 10? Ask yourself when that number is a palindrome. Here is a table you may use for the game. Start the clock now.

Number of Steps	Palindromes
0 steps	?
1 step	
2 steps	
?	?

How many two-digit numbers are on your table? If your table has all 90 two-digit numbers on it, then you have demonstrated the following fact.

> Every two-digit number is a k-step palindrome for some k.

In fact, every two-digit number is a k-step palindrome for some $k \leq 24$. To tally your score, for each k from 0 to 24, multiply the number of k-step palindromes on your table by k. Your total score is the sum of these numbers.

Here is a list of the number of k-step two-digit palindromes for $k = 0, 1, ..., 24$.

Step Number	Number of Two-Digit Palindromes
0	9
1	49
2	20
3	4
4	4
6	2
24	2

This means that the best possible score is $49 + (2)(20) + (3)(4) + (4)(4) + (6)(2) + (24)(2) = 177$.

If you scored 150 points or better, you attained the **palindrome zone.** As you work through the Post-Game Analysis below, you will be able to complete more of your palindrome table. A score of 150 points or better will be the goal in your second attempt.

Post-Game: Play-by-Play Analysis

This is a workout in base 10 arithmetic. The payoff will be a better understanding of digits, carrying and regrouping, and some new ideas for other palindrome projects. The analysis comes in the form of questions.

Each such question is marked with the ☖ symbol to remind you to stop reading and take time to think about the question. (In this section, many of the questions will have answers immediately following in the text.)

Let's look at the advantages to be gained by applying some digit dexterity. We write the two-digit number N in the form $N = 10t + u$, where t is the tens digit and u is the units digit. This means t is an integer that is between 1 and 9 and u is an integer that is between 0 and 9. For short, we use inequalities and write $1 \leq t \leq 9$ and $0 \leq u \leq 9$.

First, consider the benefits of studying the sum $t + u$ of the tens digit and the units digit. Try to answer these two questions:

How small can $t + u$ be? How big can $t + u$ be?

If your answer to the first question is 1 and to the second is 18, that's correct. Did you think about it this way? Since t is at least 1, $t + u$ is at least 1. Since both t and u are no bigger than 9, $t + u$ cannot be any bigger than 18.

Here is a follow-up question: If $t + u = 18$, what is N?

Did you see that if $t + u = 18$, then both t and u are equal to 9 and $10t + u = 99$, a palindrome? Good work.

Filling in the zero-step palindromes (that is, the palindromes) in the table is easy. So let's assume from now on that N is not a palindrome. This means that $N = 10t + u$ and that the tens digit t and the units digit u are not equal, $t \neq u$.

Here's another question. If $t \neq u$, how large can $t + u$ be?

Did you see that $t + u$ cannot be larger than 17?

Let's summarize what you have found. The sum $t + u$ satisfies $1 \leq t + u \leq 18$. If $t + u = 18$, then $N = 99$. If $t \neq u$, $1 \leq t + u \leq 17$.

Since we are looking for schemes that inform us generally about two-digit numbers, analysis of the number that is obtained at the first step is crucial.

Let's denote that number by N_1, and call it the *first-step number*. This means that if we start with the number N, and if N' is the number obtained by reversing digits, then $N_1 = N + N'$.

Good Idea 1. Find the digits of the first-step number.

Start with the number $N = 10t + u$, with $t \neq u$. Then $N' = 10u + t$ and the first-step number is

$$N_1 = N + N' = 10(t + u) + (t + u).$$

Be careful. This may not be the final form in base 10. There are two cases: either $t + u \leq 9$ or $t + u \geq 10$.

Case 1. If $t + u \leq 9$, then this is an easy case. Since $t + u \leq 9$, $t + u$ is the units digit of N_1, and since $t + u \geq 1$, $t + u$ is the tens digit of N_1. So N_1, the first-step number, is a two-digit palindrome!

Case 2. If $t + u \geq 10$, then the arithmetic is harder because carrying or regrouping is necessary. In this case, the units digit of N_1 is $t + u - 10$ and we carry 1 over to the tens place. Then, since $t + u + 1 \geq 10$, the new tens digit is $t + u + 1 - 10 = t + u - 9$. Now we carry the 1 over to the hundreds place. This gives us a three-digit number of the form

$$N_1 = 100(1) + 10(t + u - 9) + (t + u - 10).$$

So in Case 2, the first-step number is a three-digit number with hundreds digit equal to 1, tens digit equal to $t+u-9$ and units digit equal to $t+u-10$.

Let's look at some examples. Consider the Case 2 numbers 39 and 68.

Question 12. What are the digits of the first-step number for 39? How about for 68?

We will return to Case 2 numbers momentarily. First, we note that Case 1, above, answers the "what step palindrome" question for lots of two-digit numbers.

Payoff from Good Idea 1. If $t \neq u$ and $t + u$ is no bigger than 9, then N is a one-step palindrome.

Good Idea 2. All two-digit numbers with $t \neq u$ that have the same sum of digits $t + u$ are the same step palindrome.

You know that $N_1 = N + N' = 10(t + u) + (t + u)$. Even though this might not be the final base 10 form for N_1, you can see from this expression that all two-digit numbers having the same sum of digits $t+u$ have the same first-step number.

For example, numbers with $t \neq u$ having the same sum of digits $t+u = 15$ are $69, 96, 78$, and 87.

So all two-digit numbers with $t \neq u$ that have the same sum of digits $t + u$ have the same first step. But each succeeding step uses only the step before. So all two-digit numbers, with $t \neq u$ and with the same sum of digits $t + u$, will have all steps the same, and consequently will be the same step palindrome. Thus, for example, $69, 96, 78$, and 87 are all the same step palindrome.

Payoff from Good Idea 2. Assume that $t \neq u$. Fix a value of $t + u$ (between 1 and 17). Let S be the set of all two-digit numbers with that value for the sum of its digits. To determine what step palindrome all numbers in S are, it is sufficient to test any one number in S.

Since N and N' have the same sum of digits, Good Idea 2 generalizes the fact you observed earlier, namely, that, if $u \neq 0$, then N and N' are the same step palindrome.

Challenge 1. Use these ideas to make up a table of all the one-step palindromes which satisfy Case 1, namely all the two-digit numbers $10t + u$ with $t \neq u$ and $t + u \leq 9$. Here's a way to organize that list: Give t values from 1 to 9 and find all the u with $t \neq u$ and $t + u \leq 9$.

See if you can complete the following table. There are some hints to get started.

Value of t	Values of u, $t \neq u$ so that $t + u \leq 9$	Number of N with $N = 10t + u$	Values of N
$t = 1$	$u = 0, 2, 3, 4, 5, 6, 7, 8$	8	$10, 12, 13, 14, 15, 16, 17, 18$
$t = 2$	$u = 0, 1, 3, 4, 5, 6, 7$		
$t = 3$	$u = 0, 1, 2, 4, 5, 6$		
$t = 4$	$u = 0, 1, 2, 3, 5$		
$t = 5$	$u =$		
$t = 6$	$u =$		
$t = 7$	$u =$		
$t = 8$	$u =$		
$t = 9$	$u =$		

Two-digit one-step palindromes in Case 1.

It is curious, isn't it, that what appears to be a pattern of decreasing numbers $8, 7, 6, 5$ in the third column of the table has a blip at $t = 5$? Mathematical appearances can be deceiving!

The table shows that there are 41 two-digit one-step palindromes $10t + u$ with $t \neq u$ and $t + u \leq 9$. That's almost half of all of the two-digit numbers! Are these all of the one-step palindromes?

No. Earlier, you showed that 56 is a one-step palindrome, but the sum of the digits of 56 is 11 so there are more one-step palindromes to be accounted for.

Before we look for them, let's plan a strategy.

Strategy Session

There are 9 two-digit zero-step palindromes. You have found 41 two-digit one-step palindromes. Consequently, 50 of the 90 two-digit numbers, including the special numbers with units digit equal to 0, have been shown to be either zero-step or one-step palindromes. There are 40 numbers to go.

Question 13. You know that means there really are only at most twenty numbers left to test, don't you? Why?

While it is possible to finish the game by checking the remaining 20 paired numbers one pair at a time, the tactic we are going to use here is additional analysis of N_1. The bonus will be a very swift classification of 16 more numbers.

Let's go back to the one-step number in Case 2 where $t + u \geq 10$. Remember that in this case, N_1 is the three-digit number:

$$N_1 = 100(1) + 10(t + u - 9) + (t + u - 10).$$

N_1 has hundreds digit 1, tens digit $(t + u - 9)$, and units digit $(t + u - 10.)$ What does it take for N_1 to be a palindrome?

Did you see that there is only one condition for N_1 to be a palindrome, namely that the units digit of N_1 must be 1? This means that $t + u - 10$ must be 1, that is, $t + u$ must be 11.

Question 14. What are the two-digit numbers with $t + u = 11$?

Your analysis gives you eight new numbers to add to the table. But your work, combined with Good Idea 1 and Good Idea 2, actually gives much more than that. It furnishes a necessary and sufficient condition for a two-digit number N to be a one-step palindrome:

> N is a one-step palindrome if and only if
> $t + u \leq 9$ and $t \neq u$, or $t + u = 11$.

At this point, you know all about two-digit numbers with sum of digits no greater than 9, and those with sum of digits equal to 11. What happens when the sum of digits $t + u$ is equal to 10? Well, 55 is one, and it is a palindrome. That's easy. What about the numbers that have sum of digits equal to 10 and $t \neq u$? Don't check them one at a time. Use some digit dexterity and substitute 10 for $t + u$ in the first step number N_1:

$$\begin{aligned}
N_1 &= 100(1) + 10(t + u - 9) + (t + u - 10) \\
&= 100(1) + 10(1) + (0) = 110.
\end{aligned}$$

Now calculate the second step number and see what you find.

Did you do it this way? Reverse the digits to get $N_1' = 011$ so when you add N_1 and N_1' to get the second-step number N_2, you get a palindrome!

$$N_1 + N_1' = N_2 = 110 + 011 = 121$$

This shows that all two-digit numbers N with $t \neq u$ and $t + u = 10$ are two-step palindromes.

Question 15. Write down the numbers with $t \neq u$ and $t + u = 10$.

You have eight more numbers to add to the table. Let's take a moment to see how close you are to the goal.

You have all the zero-step and one-step two-digit palindromes. There are 58 of them. In addition, you have a start on the two-step palindromes; you have 8 of them. This means there are just $90 - 66 = 24$ two-digit numbers left to test. In fact, as you know, only half that many need to be tested, so only 12 are left. But, by Good Idea 2, there are even fewer to check. In fact, if you group all the numbers with the same sum of digits together and use Good Idea 2, there are only 6 more numbers to check.

The strategy now will be to finish up the game quickly (hooray!) and classify these numbers individually. Here are the untested numbers, grouped according to the sum of their digits:

39, 48, 57	49, 58, 67	59, 68	69, 78	79	89
93, 84, 75	94, 85, 76	95, 86	96, 87	97	98

Test the six numbers needed to finish the game.

When you finish the game, compare your table with the table of all two-digit numbers in the Solutions and tally your final score. Did you reach the palindrome zone (150 points or higher)? Think how rapidly you would have completed the table of two-digit palindromes on your first attempt had you understood then what you do now! You can put these strategies to good use in the challenging activities in the Heavy Lifting section.

Are you curious about three-digit numbers? Here's a surprise.

> It is not known whether every three-digit number
> is a k-step palindrome for some k or not.

Over 2,000,000 steps have been computed for the number 196 without producing a palindrome. If this intrigues you, turn to the Heavy Lifting section.

Solutions

Digit Dexterity Question. If $N = 10t + u$, with $1 \leq t \leq 9$ and $1 \leq u \leq 9$, then $N' = 10u + t$. What happens if $u = 0$?

Question 12. For 39, $t + u = 12$. The hundreds digit of N_1 is 1, the tens digit is $3 = t + u - 9$, and the units digit is $2 = t + u - 10$. So $N_1 = 132 = 1 \times 10^2 + 3 \times 10 + 2$. For 68, $t + u = 14$. The hundreds digit of N_1 is 1, the tens digit is $5 = t + u - 9$, and the units digit is $4 = t + u - 10$. So $N_1 = 154 = 1 \times 10^2 + 5 \times 10 + 4$.

Challenge 1. Two-digit one-step palindromes in Case 1.

Value of t	Values of u,u $\neq$ t so that $t + u \leq 9$	Number of N with N = 10t + u	Values of N
$t = 1$	$u = 0, 2, 3, 4, 5, 6, 7, 8$	8	$10, 12, 13, 14, 15, 16, 17, 18$
$t = 2$	$u = 0, 1, 3, 4, 5, 6, 7$	7	$20, 21, 23, 24, 25, 26, 27$
$t = 3$	$u = 0, 1, 2, 3, 5$	6	$30, 31, 32, 34, 35, 36$
$t = 4$	$u = 0, 1, 2, 3, 5$	5	$40, 41, 42, 43, 45$
$t = 5$	$u = 0, 1, 2, 3, 4$	5	$50, 51, 52, 53, 54$
$t = 6$	$u = 0, 1, 3$	4	$60, 61, 62, 63$
$t = 7$	$u = 0, 1, 2$	3	$70, 71, 72$
$t = 8$	$u = 0, 1$	2	$80, 81$
$t = 9$	$u = 0$	1	90

Question 13. Since none of the remaining 40 numbers have units digit equal to 0, N and N' are the same step palindrome. If you test half of the remaining 40 numbers, you will also know what step palindrome the numbers in the other half, with digits reversed, are. You can use Good Idea 2 to reduce the number of cases further.

Question 14. There are 8 such numbers: $29, 38, 47, 56, 65, 74, 83,$ and 92.

Question 15. There are 8 such numbers: $19, 28, 37, 46, 64, 73, 82,$ and 91.

Here is a complete table of all two-digit k-step palindromes.

zero-step palindromes		$11, 22, 33, 44, 55, 66, 77, 88, 99$
one-step palindromes	$t + u \leq 9$	$10, 12, 13, 14, 15, 16, 17, 18, 20$ $21, 23, 24, 25, 26, 27, 30, 31, 32$ $34, 35, 36, 40, 41, 42, 43, 45, 50$ $51, 52, 53, 54, 60, 61, 62, 63, 70$ $71, 72, 80, 81, 90$
one-step palindromes	$t + u = 11$	$29, 38, 47, 56, 65, 74, 83, 92$
two-step palindromes	$t + u = 10$	$19, 28, 37, 46, 64, 73, 82, 91$
two-step palindromes	$t + u = 12$	$39, 48, 57, 75, 84, 93$
two-step palindromes	$t + u = 13$	$49, 58, 67, 76, 85, 94$
three-step palindromes	$t + u = 14$	$59, 68, 86, 95$
four-step palindromes	$t + u = 15$	$69, 78, 87, 96$
six-step palindromes	$t + u = 16$	$79, 97$
24-step palindromes	$t + u = 17$	$89, 98$

Part IV: Heavy Lifting

There are three activities in this section. Each of them leads to further investigations. Let your interest be your guide. The first two projects encourage you to push to the limit with base 10 analysis of the second-step number N_2 in the two-digit case, and of the first-step number N_1 in the three-digit case. If you have had enough of all this base 10 analysis for a while, skip to the last activity, and experiment with three-digit palindromes.

You found the twelve two-step palindromes with sum of digits equal to 12 and 13 by computation. The first project is a guide for those interested in using base 10 arithmetic to find these numbers.

Project 1: Two-Digit Two-Step Palindromes

The goal is to discover when N_2 is a palindrome. You will do most of this analysis on your own, but here are some suggestions to get off to a good start. Begin, as usual, with the number $N = 10t + u$ with $1 \leq t \leq 9$ and $0 \leq u \leq 9$, and assume N and N_1 are not palindromes. You know this means that $t \neq u$, $t + u \geq 10$ (so you are in Case 2) and $t + u \neq 11$. Here's what the first-step number looks like under these conditions:

$$N_1 = 100(1) + 10(t + u - 9) + (t + u - 10).$$

To make the computation easier to follow, set $t + u - 10 = u_1$, so that $t + u - 9$, the tens digit of N_1, satisfies $t + u - 9 = u_1 + 1$. If $u_1 = 0$, then $t + u = 10$, and $N_1 = 110$. So $N_1' = 11$ has fewer digits than N_1, but just as in the previous encounter with numbers with units digit equal to 0, the reverse and add rule still can be applied. In this case, you obtain $N_2 = 121$, a palindrome.

With the case $t + u = 10$ settled, and with the assumption that $t + u \neq 11$, you may assume that $t + u \geq 12$, so that $2 \leq u_1 \leq 7$. As a result, the base 10 representation of N_1 is

$$N_1 = 100(1) + 10(u_1 + 1) + u_1.$$

Now reverse digits to get

$$N_1' = 100u_1 + 10(u_1 + 1) + 1.$$

Consequently, the second step number N_2 is:

$$N_2 = N_1 + N_1' = 100(u_1 + 1) + 10(2)(u_1 + 1) + (u_1 + 1).$$

Be careful. This may not be the final form in base 10.

Challenge 1. Assume $2 \leq u_1 \leq 7$. Find the conditions for N_2 to be a palindrome.

Hint: Is $u_1 + 1$ the units digit of N_2? Is it the hundreds digit of N_2? If not, when will it be the hundreds digit of N_2?

Challenge 2. Make a table showing all the two-digit numbers with $t + u = 10$, all with $t + u = 12$, and all with $t + u = 13$.

Solutions

Challenge 1. Did you think about it this way? $u_1 + 1$ is the units digit of N_2, since u_1 is no more than 7. For N_2 to be a palindrome, $u_1 + 1$ must also be the hundreds digit of N_2. But this will only happen if the tens digit of N_1, $2(u_1 + 1) \leq 9$. This will only happen if $u_1 \leq 3$. For if $2(u_1 + 1) > 9$, that is if $u_1 > 3$, then the hundreds digit of N_2 will be $u_1 + 2$ and will not be equal to the units digit $u_1 + 1$. This yields all the two-step palindromes. They are exactly the numbers with $u_1 = 0, 2$, and 3, that is, the numbers with $t \neq u$ and $t + u = 10, 12$, and 13. (Recall that $t + u = 11$ gives one-step palindromes.)

Challenge 2. Did you obtain these 20 numbers?

Value of t + u	Two-Step Palindromes
10	$19, 28, 37, 46, 64, 73, 82, 91$
12	$39, 48, 57, 75, 84, 93$
13	$49, 58, 67, 76, 85, 94$

Suggestions for the Endurance Athlete

10K Challenge. Collaborate with some classmates on the project of applying base 10 analysis to find all two-digit three-step palindromes. Consider the challenge of classifying all two-digit k-step palindromes in this way for $k > 3$.

Project 2: One-Step Three-Digit Palindromes.

You found all the zero-step three-digit palindromes earlier. There are 90 of them. Are you ready to take on the one-steppers? We will assume that N is not a palindrome.

This time, the number N written in base 10 has the form:

$$N = 100h + 10t + u \text{ with } 1 \le h \le 9, 0 \le t \le 9, 0 \le u \le 9 \text{ and } h \ne u$$
$$N' = 100u + 10t + h$$
$$N_1 = N + N' = 100(h + u) + 10(2t) + (h + u).$$

Again, be careful. This may not be the final form in base 10.

There are several cases to consider. Here they are.

Case 1. $h + u \le 9$ and $2t \le 9$.
Case 2. $h + u \le 9$ and $2t > 9$.
Case 3. $2t < 9$ and $10 \le h + u \le 17$.
Case 4. $2t \ge 10$ and $10 \le h + u \le 17$.

Challenge 3. Your challenge is to take each case in turn, and figure out when N_1 is a palindrome. Then write down the three-digit one-step palindromes for each case.

Solutions

Challenge 3. Did you discover that in Case 1, N_1 is always a palindrome? Good work. When you studied how many three-digit numbers satisfy these two conditions, most of the computation should have seemed familiar. There are 41 such pairs. For each of these, the tens digit t of N_1 can take on the values $0, 1, 2, 3, 4$. So there are $5 \times 41 = 205$ three-digit one-step palindromes satisfying the conditions $h + u \le 9$ and $2t \le 9$. To write them down, look back at the numbers in the second column and second row of the table you made earlier of all k-step two-digit palindromes, and insert, in turn, $0, 1, 2, 3$, and 4 as middle digit.

Did you see that, in Case 2, N_1 is never a palindrome? Good.

To analyze Case 3 and Case 4, you can take the digit sums $h + u = 10, 11, \ldots, 17$ individually and check each one, or you can write $h + u = 10 + (h + u - 10)$, substitute, regroup, and see what happens. You will find exactly eight more one-steppers: 209, 308, 407, 506, 605,704, 803, and 902.

In Case 3, for N_1 to be a palindrome, $h + u$ must be equal to 11 and t must be equal to 0. It follows that $N_1 = 1111$, so all $N = 100h + 10(0) + u$ with $h + u = 11$ are one-step palindromes.

In Case 4, for N_1 to be a palindrome, $h + u$ must again be 11. Then, for the middle digits of N_1 to be equal, you must have $2t = 11$ which is impossible. So Case 4 produces no one-step palindromes.

Suggestions for the Endurance Athlete

10K Challenge. Collaborate with some classmates to analyze N_2 for three-digit numbers as we did for two-digit numbers. Then, if you enjoy this challenge, try N_3 for three-digit numbers!

Project 3: Three-Digit k-Step Palindromes

It is not known whether every three-digit number is a k-step palindrome for some k or not. The number 196 is the smallest number which is not known to be a k-step palindrome for some k. Over 2,000,000 steps have been computed for 196 without producing a palindrome.

Begin by experimenting with three-digit k-step palindromes. Take some three-digit numbers that are less than the troublesome 196, and see what step palindrome they are. Do you see any patterns? Now go beyond 196. Are there some types of three-digit numbers that are quick and easy to characterize as one-step or two-step palindromes?

Suggestions for the Endurance Athlete

10K Challenge. Write a computer program to find the k-step palindromes among three-digit numbers.

20K Challenge. Find the smallest integer, of whatever number of digits, that is a k-step palindrome for $k = 0, 1, 2, 3, 4, \ldots$.

30K Challenge. Investigate the prime numbers that are palindromes. Show that, with the exception of the number 11, every prime number that is a palindrome has an odd number of digits.

40K Challenge. Do a Web search on number palindromes. Find a fact, not discussed here, that you think is interesting, and investigate it.

References

Eric Weisstein's *World of Mathematics* (http://mathworld.wolfram.com) is an excellent source. There you will find several directly related articles as well as many references with very timely information.

The Math Forum (www.mathforum.org) is also an excellent reference.

Cut The Knot (www.cut-the-knot.com) posts a list of many mathematical sites of interest (www.cut-the-knot.com/collection.html).

Chapter Six

The Four Numbers Game

This "mathematical curiosity" involves much more interesting mathematics than appears at first sight. You will discover that The Four Numbers Game played with positive integers comes to an end. The challenge is: Can you prove it?

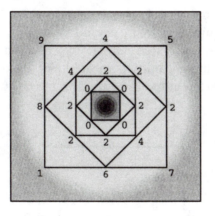

"Mathematics is a game played according to certain simple rules..."

D. Hilbert

The Four Numbers Game promises some fun, some surprises, and some very interesting mathematics. The only skills required to play the game are knowledge of subtraction and ordering of numbers. The game uses this basic arithmetic to form a sequence of smaller and smaller numbered squares, one inside the other. The big question is "Does the game terminate?"

Part I: How to Play the Game

To set up the game draw a square, the start square, and put four non-negative numbers at the corners (or vertices) of the square as pictured below.

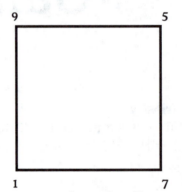

In Round 1 of the game, a second numbered square is created inside the start square. To do this, first mark the midpoints of the sides of the start square. Assign numbers to these points this way: For each side, subtract the smaller corner number from the larger. So, for example, the number $4 = 9 - 5$ is placed at the midpoint of the top of the start square, $9 - 1 = 8$ on the left side, $7 - 5 = 2$ on the right side, and $7 - 1 = 6$ on the bottom.

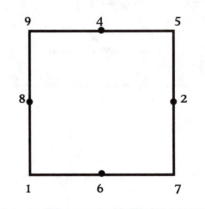

Next, connect the four new numbers to form a square inside the start square.

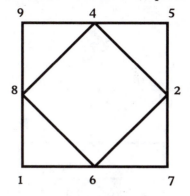

For Round 2, mark the midpoints of the sides of the second square, and apply the same procedure as in Round 1 to form a third numbered square inside the second square. Whenever two numbers at the vertices are the same, put the number 0 at the midpoint.

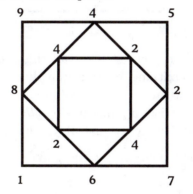

For Round 3, repeat the subtraction routine to form a fourth numbered square.

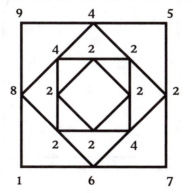

What is going to happen in Round 4 when you repeat the subtraction routine once more?

If you said a square with "all 0s," you are right. The numbers you get next are all 0s and the game ends in four rounds.

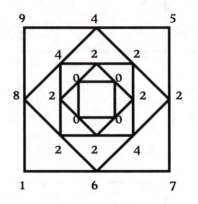

In any game, if and when you get 0s at all the midpoints, the game ends because if play were to continue, every round thereafter would still result in all 0s. We will say that a game has *length* 0 if it starts with a square having 0 at every corner. Otherwise, a game has *length* k if the first time you get all 0s is in Round k. The game we just played has length 4.

Game 1. Now it is your turn to try the game. Start with the following square.

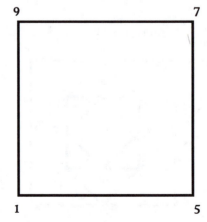

The same four numbers are used, but their starting positions have changed. See if the game ends, and if it does, how many rounds it takes.

Did you get a surprise when you played the game? The same numbers were used. But the fact that the numbers were in a different order caused this game to last seven rounds. We conclude from this that the length of the game depends not only on the numbers chosen but also on the arrangement of the numbers at the corners of the start square.

Do you think that if really big numbers are put at the corners that the game will last a long, long time?

Game 2. Try the next game and see.

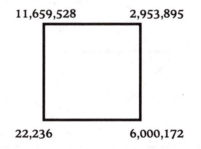

11,659,528   2,953,895

22,236   6,000,172

Don't groan, just do it!

Did the solution to this game provide another surprise? There is something interesting going on here! That means it is time to use some mathematical ideas to begin an investigation of the game.

Solutions

Game 1. The same numbers were used, and the game did end but it took seven rounds instead of four. This game has length 7.

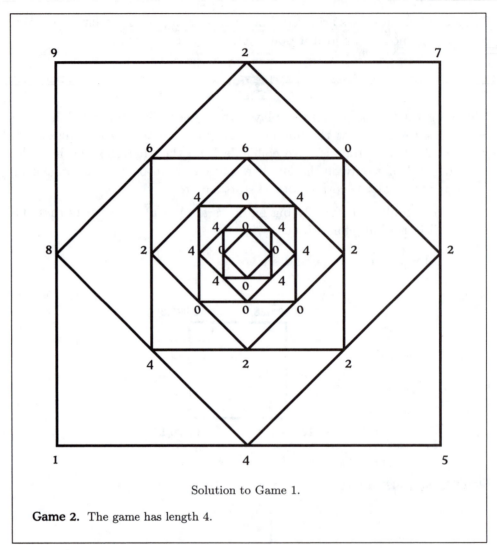

Solution to Game 1.

Game 2. The game has length 4.

Part II: Investigation of the Game

Let's look for clues to help us predict the length of a game. For the moment, we will just work with the numbers from 0 to 9.

It will be useful to have examples of very short games. There is exactly one game of length 0. That's the game that starts with a square with 0 at every corner. We will call it the *zero game*. All other games are *non-zero games*.

Here are some warm-up exercises that ask you to find some very short, non-zero games.

Warm-Up 1. Construct a game of length 1, that is, a non-zero game that ends in one round.

Warm-Up 2. Construct a game of length 2, that is, a non-zero game that ends in two rounds.

Here's your first challenge.

Challenge 1. You have examples of games of length one, two, four, and seven. Can you create a game of length three? How about a game of length six?
   Hint: Don't start from scratch here. Work with what you already have.

Drawing the squares for each round is fun, but as the games get longer, it becomes more difficult to squeeze in all the squares. Perhaps you have noticed that all that is really needed is a method for keeping a record of the four numbers at the corners of the square in each round.

Here is an example of how to do that using the very first game played.

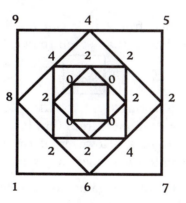

To keep a record of the numbers at the corners of the squares in each round, it is necessary to specify the order in which the numbers will be written down. We will list the numbers at the corners of the start square with the top left-hand corner first followed by the remaining three numbers moving clockwise around the square. Then, in each round, the subtraction rule is applied in the order illustrated below.

Start:	9	5	7	1
Round 1:	4	2	6	8
Round 2:	2	4	2	4
Round 3:	2	2	2	2
Round 4:	0	0	0	0

Challenge 2. Here is a table for the second game played. Try to play the game without the squares.

Start:	9	7	5	1
Round 1:	2	2	4	8
Round 2:	0			
Round 3:				
Round 4:				
Round 5:				
Round 6:				
Round 7:				

From now on, it's your choice whether you want to use the squares or not.

The next challenge invites you to find a game of length greater than seven. You can try the challenge on your own or make it a contest with some friends.

Challenge 3. Using numbers from 0 to 9, can you find four numbers that will produce a game lasting longer than seven rounds?

If you found a game lasting eight rounds, you answered the challenge. In fact, starting with four numbers from 0 to 9, every Four Numbers Game ends in eight or fewer rounds. However, if the numbers at the start are allowed to be larger than 9, then much longer games are possible.

Challenge 4. What is the longest game you can find if you start with four numbers from 0 to 44?

Hint: Be prepared to go more than 10 rounds.

Challenge 5. Try the following game.

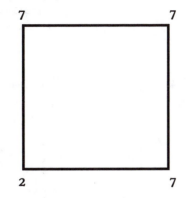

Three of the numbers are the same. It might be a fast game. Try it and see.

Do you think the number of rounds will change if you have three 8s and one 3 in place of the three 7s and one 2? How about if you have three 1,000,001s and one 563?

Challenge 6. Why don't you try the game on the following square where A and B stand for any non-negative integers, and, say, A is greater than B. To save room in the square, you might set $C = A - B$.

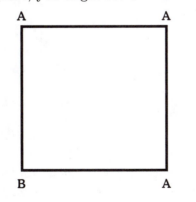

Question 1. Will the game change if B is greater than A, and you set $C = B - A$?

Algebra proves to be a very useful tool for analyzing the game. In the next challenge, you will use it to obtain information about the length of all games of a certain form. Suppose that A, B, C and D are non-negative integers with $A \geq C \geq B \geq D$.

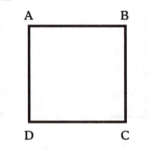

Here is how the numbers appear on a number line:

For example, if $A = 9, B = 5, C = 7$, and $D = 1$, then this is the setup of the very first game.

Challenge 7. Show that for any non-negative integers A, B, C, D with $A \geq C \geq B \geq D$, The Four Numbers Game

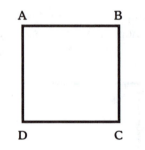

ends in four or fewer rounds.

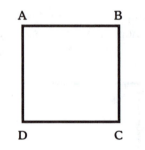

   You may get right to work on this challenge on your own, or work through it round by round with me as we read the following.
Keep in mind that $A \geq C \geq B \geq D$.

Start: A B C D

Picturing the numbers on the number line will help you use the subtraction rule to fill in Round 1.

Round 1: ? ? ? ?

Did you get this? Round 1: $A - B$ $C - B$ $C - D$ $A - D$.

The position of A, B, C, and D on the number line shows how to do the subtraction for Round 1 because it shows, for every pair, exactly which number is larger. If all four numbers A, B, C, and D happen to be equal then the game ends in one round.

In the example with $A = 9$, $B = 5$, $C = 7$, and $D = 1$, Round 1 is

$$A - B = 4 \quad C - B = 2 \quad C - D = 6 \quad A - D = 8.$$

Now for Round 2, you have to work a little harder to size up the numbers. You have $A - B$ and $C - B$. Which number is larger?

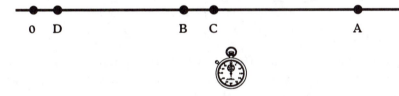

You are subtracting the same number B from both A and C, so $A - B$ is still larger than $C - B$, isn't it? (If $A = C$, then $A - B = C - B$.) For the example numbers, $A - B = 4$ and $C - B = 2$.

Next you have $C - B$ and $C - D$. Which number is larger?

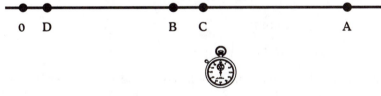

This time you are subtracting two numbers from the same number C with one number, D, smaller than the other, B. So $C - D$ will be larger than $C - B$, won't it? (If $D = B$, then $C - D = C - B$.) For the example numbers, $C - B = 2$ and $C - D = 6$.

Now it is your turn to size up $C - D$ and $A - D$, and also $A - D$ and $A - B$.

Question 2. Is $C - D$ larger or smaller than $A - D$? Is $A - D$ larger or smaller than $A - B$?

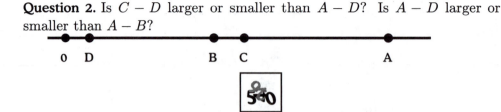

The next step is to use some algebra to do the subtractions.

Question 3. What is $(A - B) - (C - B)$?

Now you should be ready to finish the subtractions and complete Round 2.

Here are the first two rounds:

Start:	A	B	C	D
Round 1:	$A - B$	$C - B$	$C - D$	$A - D$
Round 2:	$A - C$	$B - D$	$A - C$	$B - D$

Notice that the subtractions for Round 2 give two numbers, $A - C$ and $B - D$, followed by a repeat of those two numbers. In the example with start numbers 9 5 7 1, Round 2 is: 2 4 2 4.

You have a special situation here. You have seen it before and it will likely occur in future games. Consequently, it is worthwhile to focus on it and record the result.

You have a round that looks like this:

$$\text{Round: } X \quad Y \quad X \quad Y,$$

where X and Y are non-negative numbers.

Question 4. What will the next round look like? There are several cases to tackle.

Case 1. If $X = 0$ and $Y = 0$, then there is no next round.

Case 2. If X and Y are not zero and $X = Y$, what happens at the next round?

Now suppose $X \neq Y$. Then there are two more cases, one for when X is larger than Y and one for when Y is larger than X.

Case 3. Assume $X > Y$. If you let $Z = X - Y$, what does the next round look like?

Case 4. Assume $Y > X$. If you let $Z = Y - X$, what does the next round look like?

For future reference, let's record what you have shown.

> If a game has a round of the following form
>
> *Round:* X Y X Y,
>
> and if X is not equal to Y,
> then exactly two more rounds finish the game.

Let's apply this to the A, B, C, D game, where $A \geq C \geq B \geq D$ and at least one of $A, B, C,$ and D is not equal to zero. You can think of $A - C$ as X and $B - D$ as Y.

Challenge 8. Write down all of the rounds from the start and see what you have.

Start:	A	B	C	D
Round 1:	$A - B$	$C - B$	$C - D$	$A - D$
Round 2:				
Round 3:				
Round 4:				

If you solved the A, B, C, D game correctly, then you have answered Challenge 7, and in doing so, you have proved a mathematical theorem.

Theorem 1. *For any non-negative integers A, B, C, D with $A \geq C \geq B \geq D$, The Four Numbers Game ends in four or fewer rounds.*

The next question asks you to figure out what you can say about the exact number of rounds.

Question 5. If you start with non-negative integers A, B, C, D with $A \geq C \geq B \geq D$, what conditions do A, B, C, and D have to satisfy for the game to end in zero rounds, in one round, in two rounds, in three rounds, in four rounds? Give examples of games, starting with A, B, C, D satisfying $A \geq C \geq B \geq D$, of each length from 0 to 4.

Now one final question.

Question 6. What is the length of the following game?

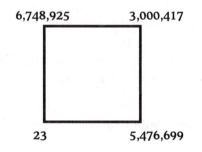

There is no need to do any computing here. Apply what you have learned.

It is true that *every* Four Numbers Game played with non-negative integers ends after a finite number of rounds.

You can investigate this surprising and interesting result in the Heavy Lifting section.

Solutions

Warm-Up 1. Every game that has the same non-zero number at all four corners of the start square ends in one round, doesn't it? Here's one with the number 4 at all four corners:

Warm-Up 2. Example of a game of length 2.

Does your game look something like this one?

Challenge 1. For a game of length 3, start the game with the square in Round 1 of the very first game in this chapter.

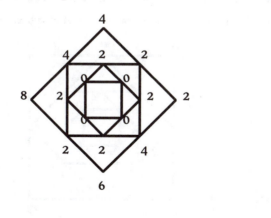

Similarly, to create a game of length 6, start with the square found in Round 1 of the second game played, the one of length 7. You can create a game of length 5 by starting with the square in Round 2 of this game, etc. You will have another example of a game of length 3 by starting with the square in Round 4.

Challenge 2. Here is the table for the second game played.

Start:	9	7	5	1
Round 1:	2	2	4	8
Round 2:	0	2	4	6
Round 3:	2	2	2	6
Round 4:	0	0	4	4
Round 5:	0	4	0	4
Round 6:	4	4	4	4
Round 7:	0	0	0	0

Challenge 3. Here is a game of length 8. Games that start with numbers between 0 and 9 all have length less than or equal to 8.

Start:	9	4	1	0
Round 1:	5	3	1	9
Round 2:	2	2	8	4
Round 3:	0	6	4	2
Round 4:	6	2	2	2
Round 5:	4	0	0	4
Round 6:	4	0	4	0
Round 7:	4	4	4	4
Round 8:	0	0	0	0

Challenge 4. What is the longest game you can find if you start with four numbers from 0 to 44? If you start with the numbers $44, 24, 13, 7$, the game lasts twelve rounds.

Challenge 5. The game lasts four rounds, a little longer than you might guess.

Challenge 6. For the game starting with $A\ A\ A\ B$, where $A \geq B$, the solution is:

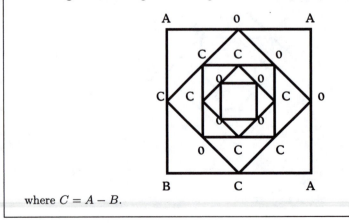

where $C = A - B$.

Question 1. No, the game and the picture are exactly the same.

Question 2. Did you see that $A - D$ is larger than $C - D$ except if $A = C$, in which case $A - D = C - D$? Did you see that $A - D$ is larger than $A - B$ except if $D = B$, in which case $A - D = A - B$? If so, that is excellent work with inequalities.

Question 3. Did you see it this way?

$$\begin{aligned}(A - B) - (C - B) &= (A - B) - C - (-B) &= (A - B) - C + B \\ &= A - B - C + B &= A - C.\end{aligned}$$

Question 4.   Round: $X \quad Y \quad X \quad Y.$

1. In Case 2, where X and Y are not zero and $X = Y$, the numbers in the next round are all zero, so the game ends.

2. In Case 3, where $X > Y$, if you let $Z = X - Y$, the next round is: $Z \; Z \; Z \; Z$.

3. In Case 4, where $Y > X$, if you let $Z = Y - X$, the next round is: $Z \; Z \; Z \; Z$.

Challenge 8. Here is the A, B, C, D game:

Start:	A	B	C	D
Round 1:	$A - B$	$C - B$	$C - D$	$A - D$
Round 2:	$X = A - C$	$Y = B - D$	X	Y
Round 3:	Z	Z	Z	Z
Round 4:	0	0	0	0

where $Z = 0$ if $X = Y$ and, if not, $Z = X - Y$ or $Y - X$ depending on which is bigger.

Question 5. (Remember that $A \geq C \geq B \geq D$.) The game ends right at the start if $A, B, C,$ and D are all equal to 0 (Not a very interesting game.). The game ends in one round if $A, B, C,$ and D are all equal to the same positive integer. The game ends in two rounds if the numbers at opposite vertices are equal: that is, if $A = C$, $B = D$, and $A \neq B$. The game ends in three rounds if $A - C = B - D$ and neither is equal to 0. To get an example of a 3-round game, you can start, say, with $A = 7, B = 3, C = 5,$ and $D = 1$. The game ends in four rounds if $A - C \neq B - D$. The game with $A = 4, B = 1, C = 2,$ and $D = 0$ has length 4.

Question 6. The game ends in exactly four rounds. Did you check to see that this is an example where $A > C > B > D$? I hope so. Since $A - C \neq B - D$, you know that the game has length 4.

Part III: Heavy Lifting

The all important fact about the Four Numbers Game played with non-negative integers is that every game ends in a finite number of rounds. You proved this for the special case where the numbers A, B, C, D satisfy $A \geq C \geq B \geq D$, and you showed that such games have length at most 4. In this "heavier" section, you will prove that *every* game has finite length. However, as you will see, the length of a game can be very large. So it will be of interest to find special cases, such as the case $A \geq C \geq B \geq D$, that force the game to end in a small number of rounds.

Rotations, Reflections, Multiplication and The Four Numbers Game

Don't you think that it might be a good idea to see if rotations and reflections (flips) are useful in the Four Numbers Game as they were in The Triangle and Square Games? It might make you a speedier and more efficient player. Here's why. Suppose you can show that rotations and reflections of the square do not change the number of rounds in the game. Then, as far as counting rounds goes, games can be considered the same if one is obtained from the other by rotations and reflections. This means that the number of cases of strictly different games will be greatly reduced.

In this part of the investigation, the squares themselves will be very helpful to our understanding of the role geometry plays in the game.

Question 7. If you rotate the square clockwise 90°, will the length of the game change? Why?

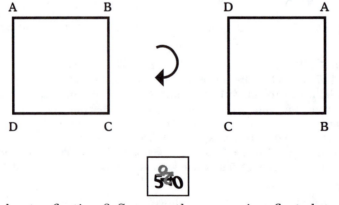

Now, how about reflections? Suppose the square is reflected across a diagonal, as pictured.

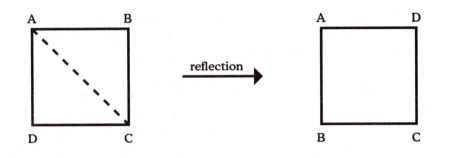

Question 8. Will the length of the game change? Why?

What's the payoff of this analysis? It is this:

> To answer questions about the length of a game,
> such as whether the game ends and how many
> rounds the game has, you may rotate or reflect
> the square as many times as you want without
> changing the answer.

There are other ways you can manipulate the start square without changing the length of the game. Here is one that involves multiplication. Try the following example.

Example 1. You know that the game with start round 9 5 7 1, in that order, has length 4. What is the length of the game that starts with 18 10 14 2? Note that the new start numbers are 2 times the old start numbers.

There is nothing special about 2. The following is true.

> Multiplication of the four start numbers
> by a positive integer does not change
> the length of the game.

The proof of this depends on the distributive law of arithmetic and on the fact that the product of two integers is zero if and only if at least one of the integers is zero. If you are curious, use these hints and try to write down a proof.

Solutions

Question 7. If you rotate the square 90°, the number of rounds does not change. Did you think about it this way? The solution of the first game is represented by a sequence of squares inside the start square. If the start square is rotated 90°, the solution square will rotate 90° at the same time, so the number of rounds will stay the same. This will hold true if the square is rotated 90°, 180°, or 270°.

Question 8. If you reflect the square across the diagonal, the length of a game will remain the same. Here's one way to think about it. If you look through the back of the paper at the square, it looks like this:

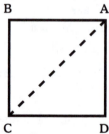

Now rotate the square 90° counterclockwise to get the reflected square.

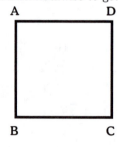

The same holds for the sequence of squares inside the start square. Thus, the reflected game has the same number of rounds as the original game. This holds true if the square is reflected across any one of the four lines pictured:

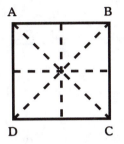

Example 1. The length of the game does not change if the start numbers are mutliplied by 2.

Start:	9	5	7	1
Round 1:	4	2	6	8
Round 2:	2	4	2	4
Round 3:	2	2	2	2
Round 4:	0	0	0	0

Here is the game with the start numbers multiplied by 2:

Start:	18	10	14	2
Round 1:	8	4	12	16
Round 2:	4	8	4	8
Round 3:	4	4	4	4
Round 4:	0	0	0	0

Predicting the Length of the Game

You showed that the length of a game does not change if the start square is rotated or reflected. This fact gives us the latitude to make some very useful assumptions about the position of the numbers at the start. For example, if you start with any four non-negative integers at the vertices of the square, and if the largest number is not at the upper left corner, you can rotate the square $90°, 180°$, or $270°$ to put it there. Then, reflecting about the NW to SE diagonal, if necessary, you can assume that the four start numbers A, B, C, D satisfy $A \geq B \geq D$ and $A \geq C$.

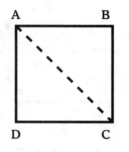

This means that the *size* of $\overset{.}{C}$ *relative to B and D is what distinguishes* the number of rounds in the game.

From now on we will always assume that the four numbers A, B, C, D satisfy $A \geq B \geq D$ and $A \geq C$ and call this the *standard form*. To make the relation of C relative to B and D precise, we name the three possibilities that can occur:

Case 1. $A \geq C \geq B \geq D$;

Case 2. $A \geq B \geq D \geq C$;

Case 3. $A \geq B \geq C \geq D$.

Your earlier work showed that Case 1 games end in four or fewer rounds. Before beginning an investigation of the remaining cases, it is useful to discuss methods for handling the subtraction rule when you are working with letters instead of concrete numbers. In particular, there is the crucial question:

> What do you do if, as you are playing the game, you do not have enough information to tell which of two numbers X and Y is larger?

There are two approaches.

1. You can split into two cases: The case X greater than or equal to Y and the case X less than Y. This may get tedious if you have to do it too often.

2. Another approach is to use the absolute value. Remember that if X and Y are two numbers, then the absolute value of $X - Y$, written

$|X - Y|$, is equal to zero if X equals Y, is equal to $X - Y$ if X is greater than Y and to $Y - X$ if Y is greater than X.

Each of the Cases 1, 2, and 3 involve possible equalities. Let us deal with all such games at once by proving the following statement.

> If any two of the numbers at the vertices of the start square are equal, then the game ends in six or fewer rounds.

If you want to try to prove this on your own, go right ahead. If you would like to work step-by-step with me, keep on reading.

Remember that the numbers A, B, C, D are in *standard form*: $A \geq B \geq D$ and $A \geq C$. So if the equal vertices happen to be A and C or B and C, you have special instances of Case 1 that you studied earlier. You know that, in Case 1, the game ends in four or fewer rounds.

Let's get started. The first two cases will assume equality of two numbers at (diagonally) opposite vertices, the remaining cases will assume equality of two numbers at adjacent vertices.

Case (i). $A = C$. Done! As remarked above, this is a special instance of Case 1.

Case (ii). $B = D$.

Challenge 9. In Case (ii) where $B = D$, write down the rounds of the game. When you don't know which of two numbers is the larger one, try using absolute value when you do the subtraction.

Your work has an interesting payoff that will be revealed by considering the following question.

Question 9. Suppose when you play the game you get to a round where you have equality of two numbers at opposite vertices. What can you say about the end of the game?

We record what you just discovered for future reference.

> If, in any round of The Four Numbers Game, two numbers at opposite vertices are equal then the game ends in four or fewer additional rounds.

Now, we return to the possible cases of equality of numbers at the vertices of the start square.

Next come the cases where the numbers at two adjacent vertices are equal.

Case (iii). $A = B$. The proof will depend on whether C is greater than or equal to D or less than D. You will see that it is worthwhile to separate the two cases $C \geq D$ and $C < D$.

Challenge 10. Suppose first that $C \geq D$. Write down as many rounds as you need to figure out the length of the game.

Hint: You may not need to play the game to the end to do this.

Challenge 11. Now assume $C < D$. Write down as many rounds as you need to determine the length of the game.

Case (iv). $\mathbf{B} = \mathbf{C}$. Done! (This is a special instance of Case 1.)

Only two more cases to go.

Challenge 12. Try to show that the game ends in six or fewer rounds for the final two cases:

Case (v). $\mathbf{C} = \mathbf{D}$.

Case (vi). $\mathbf{A} = \mathbf{D}$.

You have earned a bonus for your hard work on Cases (i)–(vi).

Bonus: From now on, you may assume strict inequality for the start numbers when you work on Cases 2 and 3.

For Case 2, the goal is to find the smallest number n so that all games with $A > B > D > C$ end in n or fewer rounds. To help you find n, try the following example. Let $A = 8$, $B = 6$, $C = 1$, and $D = 5$.

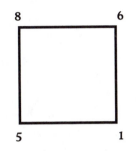

Challenge 13. What is the length of this game?

Challenge 14. The example confirms that n must be at least 6. If you can show that every game with $A > B > D > C$ has length at most 6, then $n = 6$ is the smallest such number. Try to do this now.

Your investigation has yielded many interesting facts. Here is a summary of what you have proved so far.

1. The start numbers can be put in *standard form:* $A \geq B \geq D$ and $A \geq C$.

2. If m is a positive integer, then the game starting with mA, mB, mC, mD has the same length as the game starting with A, B, C, D.

3. The game ends in four or fewer rounds in Case 1: $A \geq C \geq B \geq D$.

4. The game ends in six or fewer rounds if numbers at any two vertices are equal.

5. The game ends in six or fewer rounds in Case 2 with strict inequality: $A > B > D > C$.

To prove that every Four Numbers Game played with non-negative integers has finite length, the only case remaining is Case 3 with strict inequality: $A > B > C > D$. In this very interesting case, the game does end in a finite number of rounds, but here's a surprise.

> There is no number n with the property that every game with $A > B > C > D$ ends in n or fewer rounds.

Consequently, the proof for this case will have to be different and more abstract. In fact, the proof that you will find in Part III shows that the length of the game is finite without distinguishing cases. It will use facts 1 and 2 above, but not facts 3, 4, and 5. However, you will see that your work on Cases 1 and 2 produces much more precise information about the length of the game than the proof in the next section.

Solutions

Challenge 9. If the start square has $B = D$, here is the game:

Start:	A	B	C	B
Round 1:	$A - B$	$X = \lvert B - C\rvert$	X	$A - B$
Round 2:	$Y = \lvert A - B - X\rvert$	0	Y	0
Round 3:	Y	Y	Y	Y
Round 4:	0	0	0	0

So if $B = D$, the game ends in four or fewer rounds.

Question 9. If in a Four Numbers Game, any round has equality of two numbers at opposite vertices, then the game ends in four or fewer additional rounds. Let's review why. There is a subtle point to consider. Notice that adjacent vertices stay adjacent during rotations and flips, so opposite vertices stay opposite. This means that if you have equality of two numbers at opposite vertices then you may use rotations and flips (if necessary) to put the numbers in standard form. Now you have the standard form and two equal numbers at opposite vertices so you have either Case (i) or Case (ii).

Challenge 10 If $A = B$ and $C \geq D$, the game ends in six or fewer rounds. Let's see how it goes.

Start:	A	A	C	D
Round 1:	0	$A - C$	$C - D$	$A - D$
Round 2:	$A - C$	$\lvert A + D - 2C\rvert$	$A - C$	$A - D$

Round 2 has two equal numbers at opposite vertices so, because of your earlier work, you can conclude immediately that the game will end in four or fewer additional rounds. Consequently, the game ends in six or fewer rounds if $C \geq D$.

Challenge 11. If $A = B$ and $C < D$, the game ends in six or fewer rounds. Here is what this game looks like:

Start:	A	A	C	D
Round 1:	0	$A - C$	$D - C$	$A - D$
Round 2:	$A - C$	$A - D$	$\|A + C - 2D\|$	$A - D$

Since two numbers at opposite vertices are equal, you know that the game ends in a total of six or fewer rounds.

Challenge 12. In Case (v), when $C = D$, the first two rounds are

Start:	A	B	C	C
Round 1:	$A - B$	$B - C$	0	$A - C$
Round 2:	$\|A + C - 2B\|$	$B - C$	$A - C$	$B - C$

Once again, there are two equal numbers at opposite vertices so you know that the game ends in a total of six or fewer rounds.

In case (vi), $A = D$. You already know the answer in this case, for $A = D$ and *standard form* imply that $A = B$. You are back in Case (iii). Or, this case can be handled by observing that you have three equal start numbers. It does not matter which three of the start numbers are equal, because by employing rotations, you can see that all such games have the same length. Much earlier, you showed that a game with three of the start numbers equal has length at most 4.

Challenge 13. The game has length 6.

Challenge 14. Here are the first two rounds of the game with $A > B > D > C$.

Start:	A	B	C	D
Round 1:	$A - B$	$B - C$	$D - C$	$A - D$
Round 2:	$\|A + C - 2B\|$	$B - D$	$\|A + C - 2D\|$	$B - D$

Since Round 2 has two numbers at opposite vertices equal you know that the game ends in four or fewer additional rounds.

Proof that the Game Always Ends

The proof that The Four Numbers Game always ends when played with non-negative integers will follow from the discussion earlier in the Heavy Lifting section, and the rules of arithmetic for even and odd integers.

Here is a very interesting observation about the game.

Proposition 1. *If a Four Numbers Game has length at least 4, then all the numbers appearing from Round 4 onward are even.*

To confirm this proposition, we can play the game with the words "even" and "odd" and the rules:

$$\text{even}-\text{even} \;=\; \text{even} \qquad \text{odd}-\text{odd} \;=\; \text{even}$$
$$\text{even}-\text{odd} \;=\; \text{odd} \qquad \text{odd}-\text{even} \;=\; \text{odd}$$

The actual numbers are not needed. It is sufficient to do a case-by-case check, recording just the parity, that is, the evenness or oddness, of the four numbers at the start.

Question 10. To establish the truth of the proposition, it is enough to prove that all the numbers in Round 4 are even. Why is this so?

We need the answer to the following counting question to know how many games there are with the words "even" or "odd" at the vertices.

How many possible ways are there of arranging the words "even" or "odd" at the vertices of the start square?

There are two choices for each of the four vertices. Thus, there are $2^4 = 16$ different arrangements of "even" or "odd" at the vertices of the square. Here is one example.

Start:	even	even	even	odd
Round 1:	even	even	odd	odd
Round 2:	even	odd	even	odd
Round 3:	odd	odd	odd	odd
Round 4:	even	even	even	even

The proposition checks out in this case. Do you have to check $2^4 - 1$ more cases?

The answer is "No." You can bring rotations and reflections into the game. Rotations and reflections of the square change the position of the numbers at the vertices, but do not change the parity of these numbers. So, for example, if the proposition holds for the game that starts with "even" "even" "even" "odd," then it also holds for the game that starts with "odd" "even" "even"

"even" because the latter game is obtained by a 90° rotation. Consequently, you can verify that only six of the 2^4 start arrays need to be checked:

(i)	even	even	even	even
(ii)	even	even	even	odd
(iii)	even	even	odd	odd
(iv)	even	odd	even	odd
(v)	even	odd	odd	odd
(vi)	odd	odd	odd	odd

Notice that Cases (iii), (iv), and (vi) above occur as steps in the game that starts with Case (ii) that was just checked, so all that remains for justification of the proposition is a straightfoward inspection of Case (v).

Theorem 2. *Every Four Numbers Game with non-negative integers has finite length.*

An informal proof, with questions and examples, follows. We will need the fact that if A is any positive integer then there is a positive integer k such that A is less than 2^k.

Question 11. For example, suppose $A = 1000$. What is an example of such a positive integer k?

You may have noticed that for any positive integer A, the number A is always less than 2^A. We will also use the fact that among all such powers of 2 that are greater than A, it is true that there is a least positive power of 2 greater than A. This is an example of the use of the Well Ordering Principle for positive integers. It states that every non-empty set of positive integers has a least element.

Now we are ready to start the proof of the theorem.

Suppose that we have a Four Numbers Game with non-negative integers, at least one of which is non-zero, at the start. Suppose that A is the largest of the integers at the vertices of the start square. Let k be the least positive integer such that $A < 2^k$. We will show that the length of the game is at most $4k$.

For example, if A is a number less than or equal to 1000, then the proof will show that the game has at most $4 \times 10 = 40$ rounds because $A \leq 1000 < 2^{10} = 1024$.

If the length of the game is less than or equal to 4, there is nothing to prove. So suppose that the length of the game is greater than 4. By Propositon 1, the four numbers in Round 4 are not all zero and are even. We can create a new game with start numbers that are integers by multiplying each of the numbers in Round 4 by $\frac{1}{2}$. This new game has at least one of its start numbers not equal to zero, and has largest number at the start at most equal to $\frac{A}{2}$.

Game 3. For example, take the game that starts with 149 81 44 24. Fill in the first four rounds.

$$
\begin{array}{ll}
\text{Start:} & \quad 149 \quad 81 \quad 44 \quad 24 \\
\text{Round 1:} & \\
\text{Round 2:} & \\
\text{Round 3:} & \\
\text{Round 4:} &
\end{array}
$$

Question 12. For the above example, find the start numbers of the new game created by multiplying Round 4 by $\frac{1}{2}$. Compare the largest number of the example with the largest number of the new game.

Question 13. Here is a crucial question: For Game 3, how is the length of the original game related to the length of the new game?

Now we return to the proof where we are working with any Four Numbers Game of length greater than 4. We have created a new game with largest integer at most equal to $\frac{A}{2}$. By the multiplication property, the length of the original game is equal to the length of the new game plus 4. If the length of the new game is greater than 4, we may apply this procedure again. The four numbers in Round 4 of the new game are not all zero and are divisible by 2, so a second new game can be created with start numbers obtained by multiplying each of the numbers in Round 4 of the first new game by $\frac{1}{2}$. This second new game has at least one non-zero start number and has largest start number at most equal to $\frac{A}{2^2} = \frac{A}{4}$. The length of the original game is equal to the length of the second new game plus 8.

Before we complete the proof, return to the game created from Round 4 of Game 3, that is, the first new game that starts with 37 20 11 6.

Challenge 15. Compute the first four rounds. Find the start numbers of the second new game created by multiplying Round 4 by $\frac{1}{2}$. (This is the same as multiplying Round 8 of Game 3 by $\frac{1}{2^2} = \frac{1}{4}$.) What is the largest number of the second new game?

We return, once more, to the proof. The procedure of creating new games may be repeated each time we obtain a non-zero game after four additional rounds. Keeping in mind that $A < 2^k$, let us suppose that the length of the game is greater than $4k$. Then a new non-zero game with largest integer at most $\frac{A}{2^k}$ can be created from Round $4k$. But the non-negative number $\frac{A}{2^k}$ is less than 1. Therefore, the largest integer in Round $4k$ must be zero. So, in Round $4k$, we have all numbers equal to 0. This contradiction completes the proof of the theorem.

For Game 3 with start round 149 81 44 24, we have $149 < 2^8 = 256$. The proof shows that the length of this game is at most $4 \times 8 = 32$. In fact, the length is 15. So the bound obtained in the proof for the length of the game is much larger than the actual length.

The proof of the theorem is very interesting because, after the original estimate $A < 2^k$, it relies only on the parity of the integers. It does, however, have the drawback of providing only a very rough upper bound for the length of the game. This is why it is so useful to investigate, as you have, the special cases where the length of the game is known to be at most 4 or 6.

Solutions

Question 10. It is enough to prove that all the numbers in Round 4 are even because, thereafter in the game, every number will be obtained by subtracting two even numbers. The difference of two even numbers is even.

Question 11. If $A = 1000$, then k can be any number greater than or equal to 10, since $A = 1000 < 2^{10} = 1024$.

Game 3. Here are the first four rounds of the game that starts with 149 81 44 24.

Start:	149	81	44	24
Round 1:	68	37	20	125
Round 2:	31	17	105	57
Round 3:	14	88	48	26
Round 4:	74	40	22	12

Question 12. For Game 3, the start numbers of the new game created by multiplying Round 4 by $\frac{1}{2}$ are: 37 20 11 6. The largest number of Game 3 is 149. The largest number of the new game is 37.

Question 13. For the example, how is the length of Game 3 related to the length of the new game? The length of Game 3 is 4 plus the length of the new game. The length of the new game is 11, so the length of Game 3 is 15.

Challenge 15. Here are the first four rounds of the game created from Round 4 of Game 3.

Start:	37	20	11	6
Round 1:	17	9	5	31
Round 2:	8	4	26	14
Round 3:	4	22	12	6
Round 4:	18	10	6	2

The start numbers of the game created by multiplying Round 4 by $\frac{1}{2}$ are 9 5 3 1. The largest number at the start of this game is 9.

Suggestions for the Endurance Athlete

10K Challenge. Can you improve the estimate of $4k$ steps in the proof of the theorem that every Four Numbers Game played with non-negative integers has finite length?

20K Challenge Find another proof that the Four Numbers Game played with non-negative integers has finite length by first proving that, if $A < B < C < D$, then the largest number in Round 2 is at least one unit less than the largest integer at the start. Is the statement true if the words "Round 2" are replaced by "Round 1"?

30K Challenge. Investigate the Tribonacci numbers. The Tribonacci numbers are integers in a sequence that is defined recursively as follows. Start with $t_0 = 0$, $t_1 = 1$, $t_2 = 1$, and then for $n \geq 3$, define $t_n = t_{n-1} + t_{n-2} + t_{n-3}$. Explore the relation between the Tribonacci numbers and the length of The Four Numbers Game played with non-negative integers.

References

Freedman, Benedict, "The Four Numbers Game," *Scripta Mathematica* **14** (1948), pp. 35–47.

Webb, William A., "The Length of the Four-Number Game," *Fibonacci Quarterly* **20** (1982), pp. 33–35.

Run

Geometry

Chapter Seven

Tessellation

This investigation examines some of the geometry behind designs that are created by repetition of geometric figures. How many shapes can you find that can be fit together with no gaps or overlaps to fill up the plane?

"A mathematician, like a painter or a poet, is a maker of pattern."

G. H. Hardy

Many artistic patterns and designs have their source in geometry and exemplify a close relationship between mathematics and art. Designs that repeat one or more basic geometric figures will be the focus here. Simple examples include the patterns of a square tile floor, a brick wall, or a honeycomb. These patterns are examples of what is called a tessellation or tiling. The beauty and complexity of more elaborate tilings are illustrated by the Alhambra mosaics and the drawings of M. C. Escher.

Part I: An Introduction to Tessellation

A *tessellation*, or *tiling,* of the plane is a pattern made up of various shapes or tiles which completely covers the plane with no gaps or overlaps. Real-life examples such as the tile floor, brick wall, and honeycomb do not actually extend indefinitely, but can be imagined to do so.

The familiar pattern of a checkerboard is made by fitting together copies of a square and looks like this:

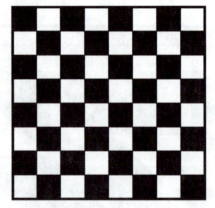

This tessellation is said to be "vertex-to-vertex and edge-to-edge." Can you see why?

A tessellation is *vertex-to-vertex and edge-to-edge* if every vertex, (an endpoint of a side), is matched only to another vertex and every edge is matched only with an edge of the same length.

Question 1. Can you find a different tessellation using copies of just one square that is not vertex-to-vertex and edge-to-edge?

The pattern of a honeycomb looks like this.

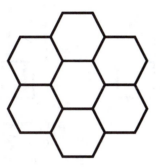

Question 2. Is it vertex-to-vertex and edge-to-edge?

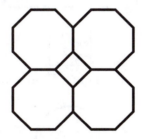

Here is another familiar vertex-to-vertex and edge-to-edge tiling using two figures, an octagon, and a square.

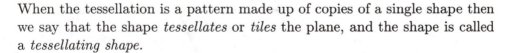

What are some other real-life examples of tessellations or tilings? Look around for some examples in which one shape is repeated and some examples in which two or more shapes are repeated.

When the tessellation is a pattern made up of copies of a single shape then we say that the shape *tessellates* or *tiles* the plane, and the shape is called a *tessellating shape*.

Tell Me if It Tessellates Game

The Tell Me if It Tessellates Game is a hunt for shapes that tessellate the plane. I will give you a shape and you must tell me whether the shape tessellates the plane or not. If your answer is "yes," you must provide a tessellation. So, for example, if I give you a square, you say "yes" and show me a tessellation such as:

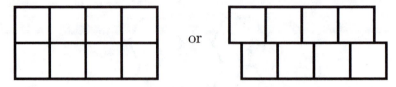

or

Answering all questions that relate to a particular class of shapes earns you a special tessellator title, beginning with Rookie Tessellator all the way to All-Star Tessellator. This game demands mathematical energy and creativity. You are going to have to work for your tessellator title.

Tessellation Game. Let's get started.

1. Does a rectangle tessellate the plane?

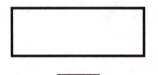

2. Does a circle tessellate the plane? Remember no overlaps or gaps are allowed.

3. Does a parallelogram tessellate the plane?

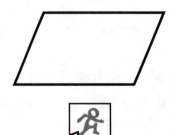

This is pretty easy so far. Did you observe that tessellating the plane with the rectangle or the parallelogram is not much different from tessellating with the square? Each of these is a shape that can be pushed to the right and left and up and down, without turning or flipping, until the plane is filled up with no gaps and no overlaps.

If a shape can be pushed to the right and left and up and down with no turns (rotations) or flips (reflections) and leaving no gaps, the shape is said to *tessellate the plane by pure translation*. This is the simplest way for a shape to tessellate. For example, a square tessellates by pure translation. For some shapes, as we shall see, rotations (turns) or reflections (flips) must be used in order to create a tessellation. Part of the strategy for these shapes will be to take two copies of the shape and rotate or flip one copy so that the two copies joined together form a shape that tessellates by pure translation.

Warning: To check whether a shape tessellates, free-hand drawings do not have the required precision. All figures must be drawn, copied, and cut out exactly. You've seen optical illusions, so you know eyes can deceive.

Does a triangle tessellate the plane?

This question is a little harder because a triangle can't just be pushed along like the rectangle and the parallelogram. Try a right triangle first.

4. Does a right triangle tessellate the plane?

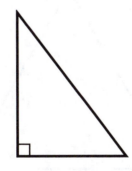

Trace and cut out a few of these right triangles. Try to fit two copies of the right triangle together along an edge in different ways until you find one that tessellates.

Here's one way of fitting two copies together along an edge.

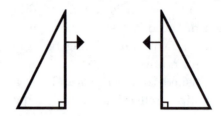

But the new figure constructed is another triangle, not something that we know yet to be a tessellating shape. See if you can find a way to fit two copies together to produce a shape that you already know is a tessellating shape.

You probably figured out by now that if copies of a new shape can be fit together to make a tessellating shape, then the new shape also tessellates the plane and is also a tessellating shape.

Did you make a rectangle out of two copies of your right triangle? If so, then, since a rectangle is a tessellating shape, you know that a right triangle tessellates the plane. Note that this tessellation by a right triangle is not by pure translation because rotations are needed.

Now try to tile the plane with equilateral triangles.

5. Does an equilateral triangle tile the plane?
Here's one you may copy.

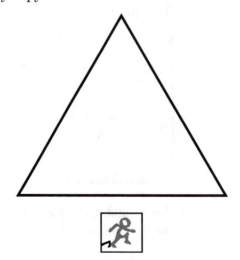

Did you fit two of these together along an edge to make a shape that you already know tessellates? What shape did you make this time?

It is a rhombus, isn't it? Recall that a rhombus is a parallelogram with all four sides equal.

Now that you have studied two special types of triangles, you should be ready to tackle the case of any triangle. Take what is called a scalene triangle.

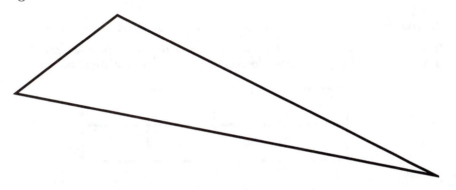

It is a triangle with no special properties at all. Use your experience with the special cases to see if you can tessellate the plane with the scalene triangle.

6. Does a scalene triangle tessellate the plane?

Trace and make some copies of the one above.

Did you see that the same idea that worked for the special cases works for any triangle? If one of the triangles is first rotated through 180° and equal sides are matched, then a parallelogram is created. Since a parallelogram is a tessellating shape, this proves that a scalene triangle is also a tessellating shape. But, since rotation was used, this tessellation does not arise by pure translation.

Congratulations! You are awarded the title of Rookie Tessellator, a title given only to those who know the following fact:

> All triangles tessellate the plane.

Solutions

Question 1. Here is an example of a tessellation that uses copies of just one square, and is not vertex-to-vertex and edge-to-edge:

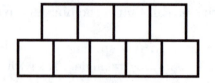

Question 2. The honeycomb tessellation is vertex-to-vertex and edge-to-edge.

Tessellation Game. For The Tell Me If It Tessellates Game, here are examples of tessellations with a rectangle, a parallelogram, and an equilateral triangle.

1. Rectangle:

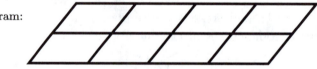

2. The circle does not tessellate the plane.

3. Parallelogram:

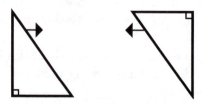

4. To show that a right triangle tessellates the plane, try to fit two copies of the right triangle together along an edge in different ways until you find one that tessellates. If one copy of the right triangle is turned or rotated and then matched with another copy this way,

a rectangle is formed. Since a rectangle tessellates by pure translation, a right triangle also tessellates the plane.

5. Equilateral triangle:

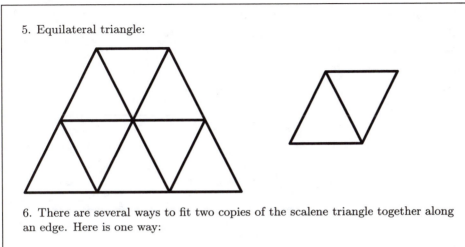

6. There are several ways to fit two copies of the scalene triangle together along an edge. Here is one way:

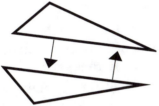

This way makes a kite, but that is not much help to you right now because you don't know yet whether a kite is a tessellating shape.

Did you make a parallelogram like this?

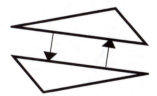

To get a parallelogram, you match the two triangles along one of the sides (it doesn't matter which side), but you have to exchange the vertices on that side. This amounts to first rotating one of the triangles through 180°, and then matching up the sides. Since a parallelogram is a tessellating shape, you have proved that a scalene triangle is also a tessellating shape.

Part II: All Quadrilaterals Tessellate the Plane

We will continue The Tell Me If It Tessellates Game. The game will culminate in the important mathematical result that all quadrilaterals tessellate the plane.

Tessellation Game (continued). You know squares, rectangles, and parallelograms tessellate. Let's continue the game with some other familiar four-sided shapes.

7. Does a trapezoid tile the plane?

Here is one to copy:

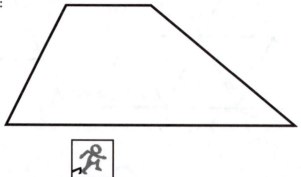

Did you find that easy? Now let's try a kite. Recall that a kite is a quadrilateral with exactly two pairs of congruent adjacent sides.

8. Does a kite tile the plane? Here is one to copy. Make about 12 copies this time. The solution for the kite will require more experimentation and ingenuity.

The kite question is not so easy. You may have tried to make a parallelogram with two copies of the kite, but it doesn't work, does it? When you match edges of two copies of the kite, you get a hexagon. Were you able to show that you can continue the pattern using your 12 copies with no gaps or overlaps? If not, try again.

If you fit two copies of the kite together to make a hexagon by first rotating one copy through 180° and then matching along a side like this:

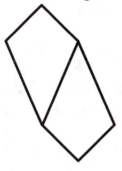

the hexagon produced has some special properties. Take a moment to see if you can detect some special features of this hexagon.

Challenge 1. Show that the hexagon constructed using two copies of the kite, by first rotating one copy through 180° and then matching along a side, has opposite sides equal and parallel.

Hint: To show that the opposite sides are parallel, try to find equal pairs of alternate interior angles. For one of the pairs of opposite sides, you will need to add a line to the picture.

Challenge 2. The next step is to show that these special hexagons tessellate the plane by pure translation. Make six special hexagons with your 12 kite copies. Arrange them carefully without gaps or overlaps to show that this special hexagon with opposite sides equal and parallel tessellates the plane by pure translation, that is, by pushing the figure in different directions without rotating or flipping.

Did you notice that to answer the "does it tessellate"? question for the four-sided kite, you showed first that the six-sided hexagon with opposite sides equal and parallel tessellates by pure translation? When you pause to think about it, you did a similar thing to show that a triangle tessellates.

You made a four-sided figure, namely a parallelogram, out of two copies of a three-sided figure, the triangle. Then you used the fact that a parallelogram tiles the plane by pure translation to show that a triangle is a tessellating shape.

The goal of this game is the proof that *all* quadrilaterals tessellate the plane. You have made good progress and have developed new skills. There is more work to do, but your experience with the kite will help you.

Here are two more quadrilaterals. These shapes represent general quadrilaterals. (The first is called a *convex scalene* quadrilateral; the second is called a *non-convex* quadrilateral. See the Heavy Lifting section for more about that.)

To attain our goal, we need to find out if the following quadrilaterals tessellate the plane.

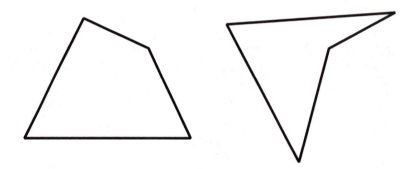

Make 12 copies of each of the quadrilaterals. For each of the shapes, see if you can make a hexagon as you did for the kite.

Some of the hexagons constructed when you fit two copies of the quadrilaterals together have better characteristics than others. For example, if you match two copies of either one of the quadrilaterals along one of the sides (it doesn't matter which side) and exchange the vertices on that side (that is, rotate one of the copies through 180°), the hexagon that you construct will have some special properties.

Question 1. Can you detect what these special characteristics are?

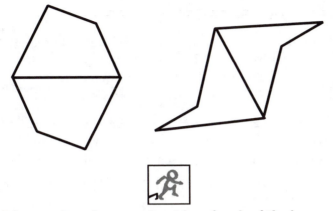

Did you observe that the opposite sides of each of the hexagons are equal and parallel? (This is the same kind of special hexagon you constructed when working with the kite.)

Challenge 3. Do the hexagons with opposite sides equal and parallel made from the general quadrilaterals tessellate the plane?

When you try to answer this hexagon question, keep in mind how the hexagons in the familiar honeycomb pattern fit together. In the honeycomb pattern, the hexagons tessellate by pure translation. You just push the hexagon around, no rotation or flipping needed, as you match up opposite sides and fill up the plane.

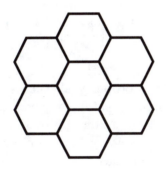

Use your copies of the quadrilaterals to make six hexagons for each of the two different quadrilaterals. For each one, construct a tessellation by fitting the six hexagon copies together, matching up opposite sides by pushing in various directions, that is, by pure translation.

Here is a summary of what you have proved about quadrilaterals so far.

> If a quadrilateral fits together with another copy
> of itself to form a hexagon with opposite sides
> equal and parallel, then that quadrilateral
> tessellates the plane.

In the process of proving this, you may have observed that all hexagons with opposite sides equal and parallel tessellate the plane.

Since we aim to prove that all quadrilaterals tessellate the plane, an answer to the following question is crucial.

Question 2. When two copies of any quadrilateral are matched up along a side by first exchanging vertices along that side, that is, by first rotating one through 180°, is the shape constructed always a hexagon?

The fact that Question 2 has a negative answer means that there is more work to do. Perhaps all quadrilaterals that do not fit together to form special hexagons have a characteristic in common that will show that they also tessellate the plane. More examples are in order. Look for at least four examples of quadrilaterals that do not fit together to form special hexagons. Then examine them to see if they have a shared property.

Here is a description of the property all such examples must have. The quadrilaterals have two supplementary adjacent angles, that is, two adjacent angles that have measures adding up to 180°. To determine what this implies, try the following exercise. The exercise is an application of some facts you know about parallel lines.

Challenge 4. Let P be a polygon with three successive sides AB, BC, and CD.

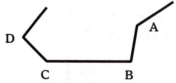

Prove the following statement and its converse. If the adjacent angles $\angle ABC$ and $\angle BCD$ are supplementary, then the sides AB and CD are parallel. Conversely, if sides AB and CD are parallel, then the angles $\angle ABC$ and $\angle BCD$ are supplementary.

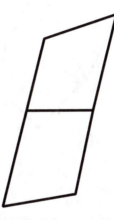

The exercise implies exactly what is needed to reach the goal of proving that all quadrilaterals tessellate the plane.

Suppose a quadrilateral has two supplementary adjacent angles. Then, if two copies are fit together, first rotating one through 180°, and then matching along the side that has supplementary angles at its vertices, a parallelogram is constructed.

Suppose the quadrilateral does not have a pair of supplementary adjacent angles. Then, if two copies are fit together, first rotating one through 180°, and then matching along a side, a hexagon with opposite sides equal and parallel is constructed.

Since you have demonstrated that parallelograms tessellate the plane and that hexagons with opposite sides equal and parallel tessellate the plane, the proof that all quadrilaterals tessellate the plane is complete.

Congratulations! You have earned the title Major League Tessellator, a title given only to those who can show the following facts:

> All triangles tessellate the plane.
> All quadrilaterals tessellate the plane.

Solutions

Tessellation Game (continued). Here is a tessellation of the plane by trapezoids:

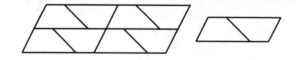

Challenge 1. Here is a hexagon constructed using two copies of the kite.

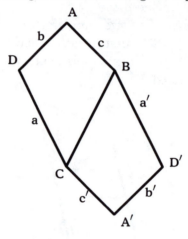

This hexagon is made by rotating one of two copies of the kite through 180° and then "gluing" along an edge. The hexagon has opposite sides equal because opposite sides of the hexagon correspond to the same side of the kite. So $a = a'$, $b = b'$, and $c = c'$. The "glued" edge is a transversal with two pairs of equal alternate interior angles: $\angle DCB = \angle D'BC$ and $\angle ABC = \angle A'CB$. It follows that a is parallel to a' and c to c'. Now add a transversal for b and b' this way:

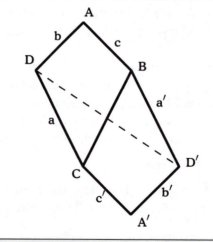

Since $\angle ADC = \angle A'D'B$ and $\angle DD'B = \angle D'DC$ (because a is parallel to a') it follows that $\angle ADD' = \angle DD'A'$, so b is parallel to b'.

Challenge 2. Does your tessellation with the special hexagons look like this?

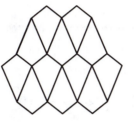

Since these special hexagons tessellate the plane, so does the kite.

Question 1. Just as in the case of the kite, the opposite sides of each of the hexagons constructed using two copies of one of the quadrilaterals are equal and parallel.

Challenge 3. Do your tessellations for the special hexagons made from the general quadrilaterals look something like these?

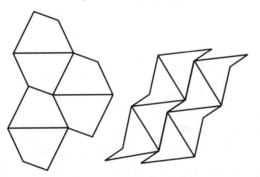

Question 2. It is not true that when you fit two copies of any quadrilateral together, first rotating one through 180° and then matching up the sides, you always get a hexagon. For example, a rectangle is formed when you match two copies of a square along a side, first rotating one through 180°.

All quadrilaterals that do not fit together after rotating one copy through 180° to form special hexagons have at least one pair of supplementary adjacent angles.

Challenge 4. Let P be a polygon with three successive sides $AB, BC,$ and CD.

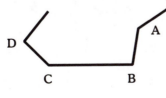

1. Suppose the adjacent angles ∠ABC and ∠BCD are supplementary. Since BC is a transversal for the sides AB and CD, it follows that AB and CD are two line segments having supplementary interior angles on the same side of a transversal. Consequently, AB and CD are parallel.

2. If the sides AB and CD are parallel, then the angles ∠ABC and ∠BCD are interior angles on the same side of the transversal CB, and must, therefore, be supplementary.

Part III: Heavy Lifting

Are you ready to become an All-Star Tessellator? To do that, you will need to recall some general facts about polygons. Here is a little review in the form of eight questions to jog your memory.

Polygon Review

Do you remember what a polygon is?

All the shapes we have been working with (except the circle) are examples of polygons, but "What is a polygon"? To draw a polygon, you make some dots (which will be the vertices), then you connect the dots to make the sides. However, there are some rules for the sides. To remind yourself what those rules are, try to answer the following question.

Question 3. Which of the following shapes are polygons and which are not?

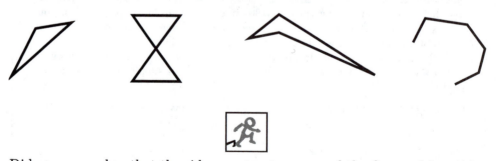

Did you remember that the sides must not cross, and the first and last sides must meet?

Here is a definition:

> Start with n distinct points $P_1, P_2, ..., P_n$ in a plane, where n is at least three, so that no three consecutive points lie on a line. If the segments $P_1P_2, P_2P_3, ..., P_{n-1}P_n, P_nP_1$ intersect only at their endpoints, they form a polygon.

Question 4. Polygons are named according to the number of their sides. How many can you name?

A polygon having n sides is called an n-gon. You could call a triangle a 3-gon if you wanted to, but the "-gon" name is most useful when the number of sides is large, or you want to make a general statement about all polygons.

Do you remember that there are two kinds of polygons, convex and non-convex?

A polygon is *convex* if the line segment joining any pair of vertices lies within the polygon. If not, the polygon is called *non-convex*. Another way to think about this idea is that a convex polygon has the measure of each of its interior angles less than 180°.

Question 5. Which of the following polygons are convex and which are non-convex?

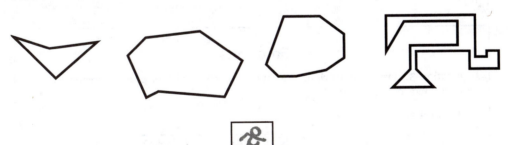

Have some fun and make a few monster-gons or nasty-gons of your own.

Question 6. Do you remember what a regular polygon is?

Question 7. What are some examples of regular polygons?

Question 8. Can you complete the table below? Recall that for any n-gon the number of degrees in the sum of all of its interior angles is $(n-2)180°$.

Number of Sides of an n-gon	Total number of Degrees in All of the Interior Angles of an n-gon	Number of Degrees in Each Interior Angle of a regular n-gon
3	180°	60°
4		90°
5		
6		
7		
8		
1000		
n		

Do you think you can sharpen your pencil to a fine enough point to draw an interior angle of a regular 1000-gon? Can you distinguish it from a straight angle?

Solutions

Question 3. From left to right, the first and the third figures are polygons; the second and fourth are not.

Question 4.

Number of Sides	Name of Polygon
3	triangle
4	quadrilateral
5	pentagon
6	hexagon
7	heptagon
8	octagon
n	n-gon

Question 5. From left to right, the first, second, and fourth polygons are non-convex; the third is convex.

Question 6. A regular polygon is a convex polygon with equal sides and equal interior angles.

Question 7. A regular triangle is an equilateral triangle. A regular quadrilateral is a square.

Question 8. Did you fill in the table below like this?

Number of Sides of an n-gon	Total Number of Degrees in All of the Interior Angles of an n-gon	Number of Degrees in Each Interior Angle of a Regular n-gon
3	$180°$	$60°$
4	$360°$	$90°$
5	$540°$	$108°$
6	$720°$	$120°$
7	$900°$	$128\frac{4}{7}°$
8	$1080°$	$135°$
1000	$179640°$	$179\frac{16}{25}°$
n	$(n-2)180°$	$\left(\frac{n-2}{n}\right)180°$

Do Pentagons Tessellate?

Let's continue The Tell Me If It Tessellates Game with some types of pentagons. A regular pentagon has five equal sides and each interior angle measures $108°$.

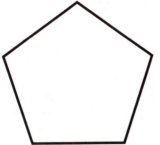

Trace and cut out some copies of this regular pentagon.

Question 9. Does a regular pentagon tessellate the plane?

First, try to answer the question for a tiling that is vertex-to-vertex and edge-to-edge. Think about how to fit the pentagons around a point in the plane.

Did you see that it is impossible to fit copies of the regular pentagon around a point? Either you get a gap or an overlap, but neither is allowed. Now, use the fact that there are 360° around every point in the plane to prove that there is no vertex-to-vertex and edge-to-edge tessellation of the plane by regular pentagons.

You have shown that there is no vertex-to-vertex and edge-to-edge tiling, but what about *any* tiling at all? Is a tiling like this possible?

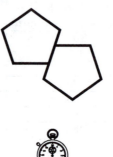

The same reasoning shows that there is no point around which an arrangement of regular pentagons will provide a sum of measures of angles equal to 360°.

Are there any pentagons that tessellate? Here are two "house" pentagons.

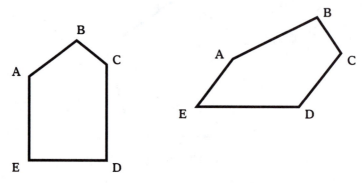

The first house has sides that meet the floor at right angles. (Consequently, it also has two opposite sides parallel.) The second "haunted" house has two opposite sides parallel.

Question 10. Do the "house" pentagons tessellate?

Cut out 12 copies of each "house" pentagon and determine whether the houses tessellate. Try the same technique of using two copies of each "house" to form a tessellating shape.

Why do the shapes you have constructed tessellate the plane?

In each pattern, angles $\angle A$, $\angle B$, and $\angle C$ fit around a point. Let us use some geometry to prove that this is so.

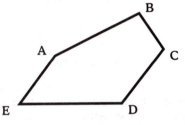

First, observe that it is sufficient to prove that the angles in each of these house pentagons satisfy $\angle D + \angle E = 180^o$. Why does $\angle D + \angle E = 180^o$ imply that $\angle A + \angle B + \angle C = 360^o$?

Next, why is it true that $\angle D + \angle E = 180^o$?

Your work shows that there is a *type* or *class* of convex pentagons that tessellates the plane. The type comprises all those convex pentagons that have two opposite sides that are parallel, or, what is the same thing, two adjacent angles that are supplementary.

Let's summarize what you have discovered about pentagons:

> There is a type of pentagon that does not tessellate the plane and a type of pentagon that does tessellate the plane. Regular pentagons do not tile, and convex pentagons with two parallel sides do tile the plane.

Are there other types of pentagons that tile the plane? The answer to this question is "yes." However, this brings us right to the edge of the unknown. The answer to the question "How many other types of pentagons tile the plane"? is not known.

At this time there are 14 known classes of convex pentagons that tessellate the plane. Some of the known tessellations are not edge-to-edge and vertex-to-vertex. The story of their discovery is full of surprises. If you are curious to learn more about this, look at the articles listed at the end of this section.

Now back to our game.

Pentagon Challenge. Construct two convex pentagons, not previously pictured, having the property that one tessellates the plane and the other does not tessellate the plane.

Solutions

Question 9. Here is a proof that there is no vertex-to-vertex tessellation of the plane by regular pentagons. Each interior angle of a regular pentagon measures $108°$. Since $3 \times 108° = 324°$, three regular pentagons around a point leave a gap of $36°$. However, $4 \times 108° = 432°$, so four regular pentagons overlap.

Question 10. If you match two copies of either one of the houses along the floors by rotating one copy through $180°$, you will get hexagons like this:

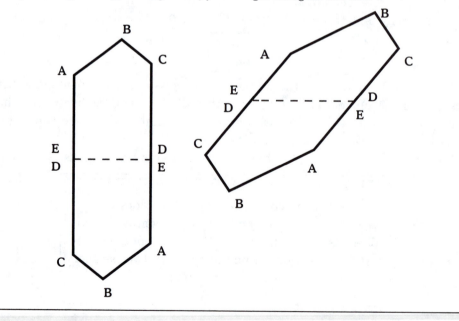

The shapes constructed with the "house" pentagons are hexagons with opposite sides that are equal and parallel. These hexagons have already been shown to tessellate the plane by pure translation. The tessellations look like this:

If $\delta + \epsilon = 180°$, then $\alpha + \beta + \gamma = 360°$ because the sum of the measures of the five interior angles of a convex pentagon is $3 \times 180° = 540°$.

The house pentagons have a pair of parallel sides joined by a third side, so the third side forms a transversal, and δ and ϵ are interior angles on the same side of a transversal. (See Challenge 4 in the section on quadrilaterals.)

What about Hexagons and Other *n*-gons?

Let's play the game with hexagons. You have a head start since you already know that hexagons with opposite sides equal and parallel tessellate the plane.

Hexagons are not as puzzling as pentagons. The question of which types of convex hexagons do or do not tile the plane is completely settled. The problem is much simpler than for pentagons because it is known that a tessellation with a convex hexagon must be edge-to-edge and vertex-to-vertex. The answer is that there are exactly three types or classes of convex hexagons that tessellate the plane. Hexagons with opposite sides equal and parallel are examples of one of the types. To find out about the other classes, look at the articles listed at the end of this section. In the meantime, play the game!

Hexagon Challenge. Construct two convex hexagons, not previously pictured, having the proprety that one tessellates the plane and the other does not tessellate the plane.

Are 7-gons up for discussion next? It may surprise you to discover that the action moves very quickly from hexagons to general n-gons. An astonishing fact will bring the tessellating shapes game to a speedy conclusion, perhaps different from what you expected. First, let's play the game with regular n-gons.

Question 11. Tell which regular n-gons tessellate the plane in a vertex-to-vertex and edge-to-edge tiling. Tell which regular n-gons do not tessellate the plane in a vertex-to-vertex and edge-to-edge tiling.

The table you made in the Polygon Review is exactly what is needed here.

Now here is a truly amazing fact.

> No convex polygons with greater than 6 sides
> tessellate the plane.

This means your observation that, for $n > 6$, regular n-gons do not tessellate the plane in a vertex-to-vertex and edge-to-edge tiling, is true for *any* convex n-gon with *any* kind of tiling.

What about non-convex n-gons?

Do there exist non-convex n-gons for n greater than 6 that tessellate the plane?

Use your ingenuity to construct a non-convex n-gon that tessellates the plane.

Here is an example of a non-convex 7-gon that tessellates the plane.

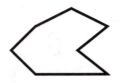

Thus the amazing fact is not true for non-convex n-gons. To construct the 7-gon above, a "bite" was taken out of one side of a tessellating house pentagon and put back on the other side.

Challenge 5. Construct a non-convex 10-gon and a non-convex 27-gon that tessellate the plane.

You might try the technique of altering a figure that you already know tessellates.

Congratulations! You have earned the title of All-Star Tessellator. This is a title given only to those who know the following facts:

> All triangles tessellate the plane.
>
> All quadrilaterals tessellate the plane.
>
> Regular pentagons do not tile, but there are at least 14 types of convex pentagons that do tile the plane.
>
> There are exactly 3 types of convex hexagons that tile the plane.
>
> No convex polygon with more than six sides can tessellate the plane.

Solutions

Question 11. Exactly three regular n-gons provide an edge-to-edge and vertex-to-vertex tessellation of the plane. The regular 3-gon (equilateral triangle), 4-gon (square), and 6-gon (regular hexagon) tile the plane vertex-to-vertex and edge-to-edge.

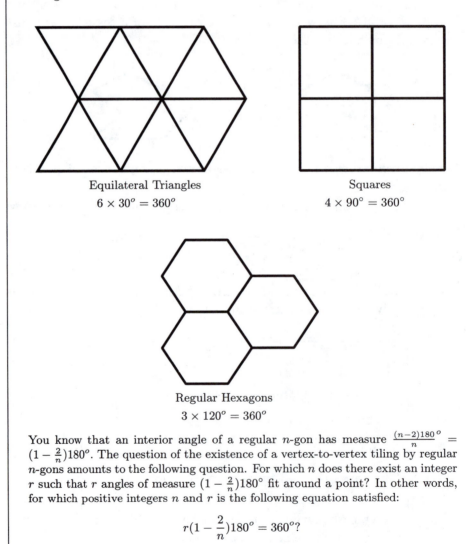

Equilateral Triangles
$6 \times 30^\circ = 360^\circ$

Squares
$4 \times 90^\circ = 360^\circ$

Regular Hexagons
$3 \times 120^\circ = 360^\circ$

You know that an interior angle of a regular n-gon has measure $\frac{(n-2)180^\circ}{n} = (1 - \frac{2}{n})180^\circ$. The question of the existence of a vertex-to-vertex tiling by regular n-gons amounts to the following question. For which n does there exist an integer r such that r angles of measure $(1 - \frac{2}{n})180^\circ$ fit around a point? In other words, for which positive integers n and r is the following equation satisfied:

$$r(1 - \frac{2}{n})180^\circ = 360^\circ?$$

This equation is the same as the equation

$$2(n + r) = nr.$$

The only pairs (n, r) of positive integers for which this equation holds are $(n, r) = (3, 6)$, $(4, 4)$, and $(6, 3)$. If you are very clever and note that the equation can be rewritten in the form

$$(n - 2)(r - 2) = 4,$$

the result is more obvious.

Challenge 4. Examples of non-convex polygons that tessellate.

Here is an example of a tessellating non-convex 10-gon. A "bite" was taken out of one side of a rectangle and put back on the opposite side.

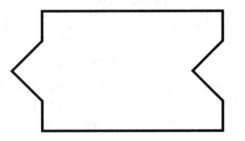

Here is an example of a tessellating 27-gon. It was made by taking "bites" out of one of the sides of a tessellating house pentagon and putting the bites back on the other side.

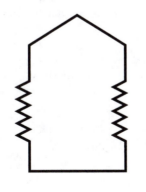

Suggestions for the Endurance Athlete

10K Challenge. Try to prove this amazing fact: No convex polygon with more than 6 sides tessellates the plane. If you and your classmates can find a proof without consulting the references, you will be inducted into the Tessellators Hall of Fame. Go for it!

20K Challenge. The dual of a tessellation by regular n-gons is formed by taking the center of each polygon and joining it to the centers of adjacent polygons. Investigate the dual of each of the three tessellations by regular n-gons.

References

Note: The references for Fulton and Niven explicitly relate to the "Amazing Fact." The articles by Gardner and Schattschneider are more expository in nature.

Fulton, C., "Tessellation," *Amer. Math. Monthly*, **99**:5 (1992) pp. 442–445.

Gardner, M., *Time Travel and Other Mathematical Bewilderments,* W. H. Freeman & Co., August, 1987.

Jacobs, H., *Geometry*, W. H. Freeman & Co, 1987.

Niven, I., "Convex Polygons That Cannot Tile the Plane," *American Mathematical Monthly*, **85** (1978), pp. 785–792.

Schattschneider, D., "In Praise of Amateurs" in *The Mathematical Gardner,* D. Klarner ed., Wadsworth International, 1981.

www.camosun.bc.ca/ jbritton/jbsymteslk.htm has links to many Web sites on symmetry and tessellation for grade 5–8 teachers and students.

www.cut-the-knot.com

www.geom.umn.edu contains an extensive and comprehensive bibliography (www.geom.umn.edu/software/tilings/TilingBibliography.html) compiled by D. Schattschneider.

www.mathforum.org

www.mathworld.wolfram.com

Chapter Eight

Circle Packing in the Plane

Since tiling the plane with circles or disks is impossible, one option is to look for arrangements of circles that cover as much of the plane as possible with no overlaps. This is the mathematical art and science of packing. To start the expedition, you will experiment with packing pennies onto squares. This will lead to some surpising discoveries about circle packing of the whole plane.

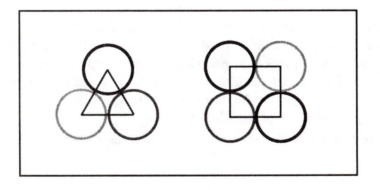

"A circle is a happy thing to be..."

C. Morley

A new set of fascinating questions arises when we focus on circles and other non-tessellating shapes in the plane. Since circles do not tile the plane, we look for arrangements of circles with the smallest possible gaps. How small can we make the gaps and still have no overlaps? We will explore this question as well as many others on the topic of packing areas of the plane. In mathematics, *packing* means filling a given space with a given set of objects allowing no overlaps. We will focus on packing areas of the plane with copies of one geometrical shape. There are two main questions to study. How many copies of the shape can be arranged with no overlaps onto a given space? How much of the space can be covered by copies of the shape?

For a real life example, suppose the supervisor for your summer job at SportsSmart asks you to arrange as many cans of new tennis balls as possible on the display shelf. Since the can bottoms are disks and disks don't tessellate, this becomes a disk or circle packing problem. When you put the cylindrical cans on the shelf, upright so the brand names show, there will be gaps between cans. You need answers to the following questions. What is the largest number of cans that can be displayed on the shelf? How much of the shelf can be covered?

Can you think of some other examples of everyday packing problems?

We will undertake four packing projects in which pennies will be substituted for the cylindrical cans. Much of what you discover about packing pennies will have practical applications. The equipment required for the projects consists of 150 pennies and three $6'' \times 6''$ squares of cardboard. The pennies are stand-ins for the cans of tennis balls and the cardboard squares represent parts of the display shelf.

Before you put your ruler away, measure the diameter of one of your pennies.

Did you get approximately $\frac{3}{4}$ inch? In the work that follows, we will assume that the diameter of one penny measures $\frac{3}{4}$ inch. (All measurements in this section are in inches; sometimes we will use the notation $''$ for inches, and occasionally the unit of measurement will be omitted in calculations.)

Part I: Packing Projects

We begin with a warm-up exercise.

Penny Packing Warm-Up

Put as many pennies as you can on one of the $6'' \times 6''$ squares with no overlaps and no out of bounds (that is, the pennies may not extend beyond the square).

Will there be gaps? Of course, because pennies don't tessellate.

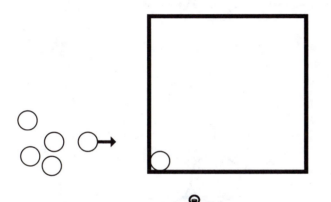

How many pennies did you pack on your square? Check out other arrangements.

For each penny packing exercise to follow, first try to estimate the largest number of pennies that can be packed on the square board. Then do the packing.

Packing Project 1

Estimate the largest number of pennies that can be packed on the square. Pack as many pennies as you can on your square with no overlaps and no out of bounds.

How many pennies did you pack? I packed 68 pennies on my square. How does the number you packed compare with 68?

Here are two very familiar examples of packing.

Square packing:

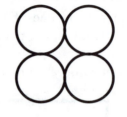

Triangle packing:

The triangle packing is sometimes called the hexagonal packing. Why do you suppose this is so?

Challenge 1. Experiment with each of these packings. See how many pennies you can pack on the square using the triangle packing, and then try the square packing.

Here is an important observation that will play a role in future exercises.

Look at a triangle packing on the square.

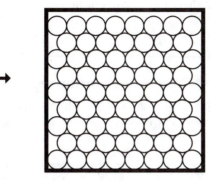

There is space between the tops of the pennies in the top row and the top of the square. Not so for the square packing.

The next part of the project is to calculate how much of the board your penny packing covers. In other words, the task is to find the following fraction:

$$\frac{\text{area of board covered by pennies}}{\text{total area of board}}.$$

This number, written as a percent, is called the *density* of the packing.

Challenge 2. Compute the density of the square packing, the triangle packing, and the packing you originally used if it was different from these two.

For some hints on how to do this, keep on reading. Otherwise, begin the computation on your own.

Hints:

1. First of all, find the area of the $6'' \times 6''$ board.

2. Next, find the total area of all the pennies. That isn't too hard to do. Remember that the formula for the area of a circle is πr^2, where r is the radius of the circle. You know the diameter of one penny is $\frac{3}{4}$ inch, so what is the radius r of the penny? Now you are set to calculate the area of one penny and then multiply by the number of pennies in your packing to compute the area covered by the pennies. An estimate of 3.14 for π is good enough.

3. Now find the ratio of the packing area to the board area. Finally, change that fraction to a percent to get the density of the packing.

Did you notice that, in the competition between the square and triangle packings, the triangle packing wins hands down? The density for the triangle packing is about 5% higher.

Solutions

Challenge 1. On the $6'' \times 6''$ board with the triangle packing, there are five rows of 8 pennies and four rows of 7 pennies for a total of 68.

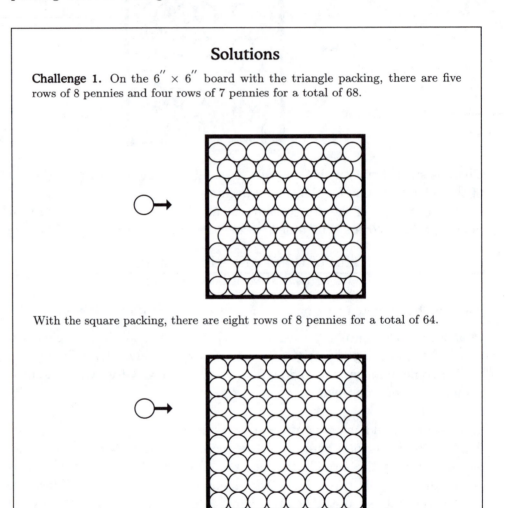

With the square packing, there are eight rows of 8 pennies for a total of 64.

Challenge 2. Here is the computation of the density of the triangle packing on the $6'' \times 6''$ board.

1. The area of the board is $6 \times 6 = 36$ square inches.

2. The diameter of one penny is $\frac{3}{4}$ inch so the radius is $\frac{3}{8}$ inch. This means the area of one penny is $\pi(\frac{3}{8})^2 = (\frac{9}{64})\pi$ square inches. Since the triangle packing uses 68 pennies, the total area of the penny packing is $68 \times (\frac{9}{64})\pi$ or approximately 30.03 square inches.

3. The triangle penny packing is approximately $\frac{30.03}{36.00}$ of the board. Changing to a percent gives a density of a little more than 83.4%. We will approximate the density by 83.4%.

To compute the density for the square packing, just change 68 to 64 in Step 2 above. For the square packing, the density is about 78.5%.

If you used a different packing in the experiment, the computation of the density of your packing should follow the steps above but the number of pennies may be different.

Packing Project 2

In Project 2, a larger rectangular board will be used. The size of a rectangle is written in the form: base × height. Build a $12'' \times 6''$ rectangular board made by placing the two $6'' \times 6''$ squares side by side with edges touching. You may use any method of packing in this project.

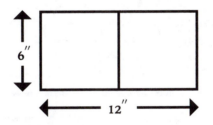

Challenge 3. Pack as many pennies as you can on the $12'' \times 6''$ board with no overlaps and no out of bounds.

Estimate the largest number of pennies that can be packed on the board. Will it double the old number? Could it possibly be smaller or larger than double the old number?

How many pennies did you pack? Were you able to do better than double your old number?

It will be interesting to see what changes, if any, there are in the densities of the various packings.

Challenge 4. Compute the density of the square and triangle packings, and of the packing you used, if it was different, on the $12'' \times 6''$ board.

Hint: To compute the density of the square packing, no computation is needed. For the triangle packing, calculation of the density follows exactly the same steps as before.

Solutions

Challenge 3. How many pennies did you pack? Is it more than double the number you got in Project 1?
With the triangle packing, the 12″ × 6″ board holds 140 pennies.

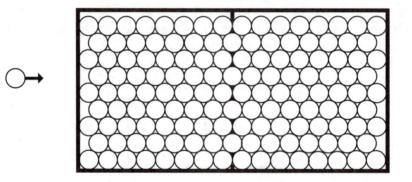

With this packing, at the middle of the 12″ × 6″ board where the squares meet, the edge gaps join up to make space for four extra pennies. So the number of pennies packed is $2 \times 68 + 4 = 140$. With the square packing, the 12″ × 6″ board holds 128 pennies, just double the number packed on the 6″ × 6″ board.

Challenge 4. The density of the square packing on the larger 12″ × 6″ board is exactly the same as before, namely, about 78.5%. The area of the board doubled and the area covered by the pennies also doubled, so the density is the same. For the triangle packing, the board holds double the number of pennies in Project 1 plus four more. The density of the triangle packing on the 12″ × 6″ board is about 85.9%. To compute it, multiply the number of pennies, 140, by the area of one penny, $(\frac{9}{64})\pi$ square inches, and divide by the area of the board, 72 square inches.

Packing Project 3

For Project 3, we will use the 6″ × 12″ board made by putting one square on top of the other with edges touching.

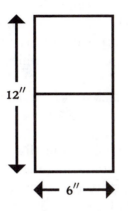

You know, by now, what will happen if you use the square packing on the $6'' \times 12''$ board, don't you?

If you said that the number will double, you are right.

In this project, you will use the triangle packing. Think before you make your estimate.

Challenge 5. Using the triangle packing (packing from bottom to top and starting with 8 pennies on the bottom row), pack as many pennies as you can on the $6'' \times 12''$ board.

Estimate the largest number of pennies that can be packed on the board.

Use the triangle packing to pack as many pennies as you can on the $6'' \times 12''$ board.

Were you surprised by the number of pennies packed in Project 3? Notice something very interesting. Remember the space you observed at the top of the board when you used the triangle packing on just one square? That space gets larger when two squares are stacked vertically. How many squares stacked vertically do you think it takes to squeeze in a whole extra row of pennies?

(In the Heavy Lifting section, you will determine exactly what this number of squares is.)

When we used the triangle packing on the $6'' \times 12''$ board, could we have done something different when we made the transition from the bottom square of the board to the top square? Is there a way to pack the new pennies so that the total number of pennies will be double the original number, $136 = 2 \times 68$, instead of one less than double, 135?

Sure there is. Since gaps are allowed in this game, a space between row 9 and row 10 is permissible. Instead of filling in the gaps at the place where the two square boards meet, start a new triangle packing with a row of eight pennies at the bottom edge of the top square:

Repetition of the triangle packing on the $6'' \times 12''$ board gives 10 rows of eight pennies and 8 rows of seven pennies, double the original number. Let's call this variation of the usual triangle packing the *repeat triangle packing*.

What do we conclude about these different packings? Which one packs the most pennies? Both the triangle packing and the repeat triangle packing beat the square packing. When the size of the board is increased horizontally to $12'' \times 6''$, with both the triangle packing and the repeat triangle packing, the board is filled with 140 pennies, compared to 128 pennies with the square packing.

When the size of the board is increased vertically to $6'' \times 12''$, a continuation of the triangle packing from the bottom square admits 135 pennies and the repeat triangle packing admits 136 pennies compared to 128 pennies with the square packing. Notice that if we take the horizontal $12'' \times 6''$ board with the triangle packing and rotate it 90°, a packing of the $6'' \times 12''$ board with 140 pennies is obtained.

Question 1. What happens when a third $6'' \times 6''$ square is added vertically?

Challenge 6. Compute the density of the triangle packing and of the repeat triangle packing on the $6'' \times 12''$ board.

Hint: Do you see that no computation is necessary for the repeat triangle packing? Why?

Solutions

Challenge 5. When you use the triangle packing on the $6'' \times 12''$ board, if you start at the bottom with a row of 8 pennies, you will have five rows of 8 pennies and four rows of 7 pennies on the bottom $6'' \times 6''$ part of the board. Now when you add the new pennies and let the first new row of 7 pennies overlap the bottom board, there is space for five rows of 7 pennies and four rows of 8 pennies.

So 67, the number of new pennies added, is one less than the number of pennies you placed on the bottom square. The total number of pennies is $(5)(8)+(4)(7)+(5)(7)+(4)(8) = 68+67 = 135$. This is one less than double the original number but still better than what you get with the square packing on the $6'' \times 12''$ board.

Question 1. When a third $6'' \times 6''$ board is added vertically in the competition between the triangle packing and the repeat triangle packing, the number of new pennies added is exactly the same for both packings, namely 68. So the total for the $6'' \times 18''$ board is $68+67+68 = 203$ for the triangle packing and $68+68+68 = 204$ for the repeat triangle packing. Notice that the space between the top of the board and the top row of pennies in the triangle packing is getting larger. For more about this, see the Heavy Lifting section.

Challenge 6. The density of the repeat triangle packing on the $6'' \times 12''$ board is exactly the same as the density of the triangle packing on the $6'' \times 6''$ board, or approximately 83.4%. The area of the board doubled and so did the area covered by pennies because the number of pennies doubled.

Here is the computation of the density for the triangle packing. The area of the board is $6 \times 12 = 72$ square inches. The area of one penny is $(\frac{9}{64})\pi$ square inches so the total area of pennies packed is $135 \times (\frac{9}{64})\pi = (\frac{1215}{64})\pi$ square inches which we estimate to be about 59.61 square inches. Consequently the penny-covered area is approximately $\frac{59.61}{72.00}$ of the board area. Changing to a percent gives a density of about 82.8%. So the repeat triangle packing has the edge here, as we know.

Packing Project 4

We return to that display shelf at SportsSmart. That shelf is $60''$ long and $18''$ deep but instead of packing it with cans of tennis balls, pennies will be used.

Challenge 7. Using either the triangle packing or the repeat triangle packing, and packing from the back of the shelf to the front, how many pennies can you pack on the display shelf at SportsSmart?

Think of the shelf as composed of thirty $6'' \times 6''$ squares arranged so that there are 10 columns of 3 squares each with each column running from the back of the shelf to the front.

Too many pennies are required here to actually do the packing, so use your experience from the first three projects and some mathematical analysis to figure it out.

Solutions

Challenge 7. Think of the shelf as composed of thirty $6'' \times 6''$ squares arranged so that there are 10 columns of 3 squares each with each column running from the back of the shelf to the front.

Triangle packing. In each column of three $6'' \times 6''$ squares, you can pack $68 + 67 + 68 = 203$ pennies. For there are $5 + 4 + 5 = 14$ rows of 8 pennies and $4 + 5 + 4 = 13$ rows of 7 pennies in each column. Now you will be able to fit in extra pennies at each column juncture: one extra penny for each row of 7 pennies. So at each column juncture, you can fit in 13 extra pennies. Thus the total number of pennies packed is $(10 \times 203) + (9 \times 13) = 2147$. (The density of this packing is about 87.8%.)

Repeat triangle packing. In each column of three $6'' \times 6''$ squares, you can pack $68 + 68 + 68 = 204$ pennies. There are $5 + 5 + 5 = 15$ rows of 8 pennies and $4 + 4 + 4 = 12$ rows of 7 pennies. You will be able to fit in extra pennies at each column juncture: one extra penny for each row of 7 pennies. So at each column juncture, you can fit in 12 extra pennies. Thus the total number of pennies packed is $(10 \times 204) + (9 \times 12) = 2148$, one more than the triangle packing! (The density of this packing is also about 87.8%.)

Suggestions for the Endurance Athlete

10K Challenge. Find a way to show that no more than 68 pennies can be packed on a $6'' \times 6''$ square.

20K Challenge. If you enjoy packing puzzles, take other non-tessellating shapes and experiment with ways to pack them. Try packing regular pentagons, regular octagons, ovals, or any other shapes you like. For a particular shape, compare the densities of various packings.

Part II: Heavy Lifting

In the first four projects, you studied what happens to the density of various packings as the size of the rectangular area increases. Here, this idea is carried to the max! The very interesting and challenging mathematics in Projects 5 and 6 will take you into more difficult territory. In Project 5, you will compute the density of packings of the whole plane. In Project 6, you will investigate the rivalry between the triangle packing and the repeat triangle packing.

Packing Project 5

Here is a warm-up exercise.

Imagine packing the whole plane with pennies. If the square packing is used, is there a way to compute the density of this packing? The plane is infinite. Is it possible to compute how much of the plane is covered by pennies?

Yes. The density is exactly the same as for the $6'' \times 6''$ square, or about 78.5%. Remember that in the square packing of the $6'' \times 6''$ square, every row and every column of pennies touches the edges of the square. So, as we saw earlier, when we fit two $6'' \times 6''$ squares together, side-by-side or one on top of the other, the number of pennies just doubles as the board area doubles. There is no room for extra pennies. This means that the density on the larger board is exactly the same as on the square. But squares tessellate the plane by pure translation, so we can fill up the plane with $6'' \times 6''$ squares just by pushing up and down and left and right.

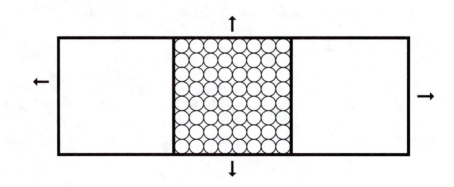

As we do this, the number of pennies increases exactly as the number of squares increases. So the density remains the same for the whole plane.

Next we investigate the density of the triangle packing on the whole plane. Imagine packing the whole plane with pennies using the triangle packing. Can the density of this packing be computed? Remember that the density of the triangle packing can increase or decrease as the area increases. Can you determine a way to use a tessellating shape to help find the density of the triangle packing on the whole plane as we did for the square packing?

Think about this for awhile and then try the computation on your own, or read on for some ideas.

To help us think about the problem and to prepare for a surprise later, we are going to draw circles larger than pennies and call the radius of the circle r. Since tessellating squares worked for the square packing, perhaps we should try to investigate the triangle packing by using tessellating triangles. Look what happens when we try to draw triangles around the circumferences of triangularly packed pennies:

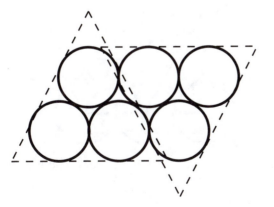

The triangles overlap. These triangles will not form part of a tessellation of the plane.

Don't give up on triangles however. The clever idea here is to use *another* triangle closely associated with the triangle packing of circles:

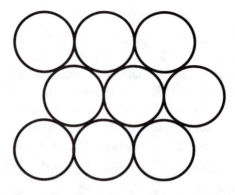

Can you find another triangle that could be used? Look for a pattern of triangles in the triangle packing that has no overlaps.

Hint: Use the centers of the circles.

Did you discover the pattern of triangles formed by connecting the centers of the circles? (Note that the fact that the line joining the centers of two tangent circles passes through the point of tangency is used to justify the formation of triangles here.)

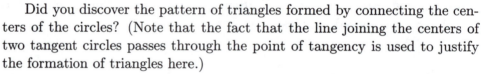

What type of triangle is formed by connecting the centers?

Can you prove it?

Did you see that the triangle formed by connecting the centers of the circles in the triangle packing is an equilateral triangle? We will call it the basic equilateral triangle. Each side has length equal to $2r$, where r is the radius of the circle.

Look again at the triangle pattern embedded in the packing and imagine it filling up the plane.

The area of the packing inside each triangle is exactly the same and consists of three congruent sectors of a circle of radius r. So, since the triangles tessellate the plane, the area of packing increases exactly as the number of triangles increases. We arrive at the following important conclusion.

> The density of the triangle packing of circles in the plane is the same as the density of the packing in one triangle.

The mathematical analysis employed here greatly simplified our problem. The original density question has been reduced to computing density in the basic equilateral triangle. That is a much more manageable question. You have all the tools required to compute the density of the packing in the triangle on your own. If you wish, you may follow the step-by-step outline given below.

Challenge 7. Compute the area of the basic equilateral triangle with side of length $2r$.

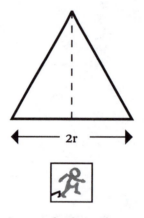

Challenge 8. The next step is to calculate the area of the packing inside the basic equilateral triangle. The packing area is made up of three congruent sectors of circles.

Each sector

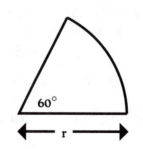

has central angle equal to $60°$ and radius r. You need to find the area of three sectors.

Challenge 9. To finish the density computation, divide the total area of the three sectors by the area of the triangle and change to a percent.

Some magic happens! In the density computation

$$\frac{(\frac{1}{2})\pi r^2}{\sqrt{3}r^2} = \frac{\pi r^2}{2\sqrt{3}r^2} = \frac{\pi}{2\sqrt{3}},$$

the factor r^2 appears in both the numerator and the denominator of the first two fractions, and since $r^2 \neq 0$, it may be cancelled for the final simplification.

What does the "disappearance" of r in the density fraction mean?

It signifies something very interesting. Namely that the circle used for packing can have a huge radius or a tiny radius, it doesn't matter because the density remains the same, about 90.7%. We have the following very surprising fact.

> The density of the triangle packing of circles in the plane does not depend on the size of the circles but only on the method of packing.

Notice that starting from the $6'' \times 6''$ board, as you pack larger and larger areas, the density gets closer and closer to 90.7%. There's a good question to ask here. See the Suggestions for the Endurance Athlete at the end of this chapter.

Solutions

Challenge 7. The area of an equilateral triangle with side of length $2r$ is $\sqrt{3}r^2$. For the height is $\sqrt{3}r$, so the area is $\frac{1}{2}(2r)(\sqrt{3}r) = \sqrt{3}r^2$.

Challenge 8. Each sector is equal to $\frac{1}{6}$th of a circle of radius r. Consequently, the three sectors together comprise half a circle, so the area of the three sectors inside the basic equilateral triangle is $\frac{1}{2}\pi r^2$.

Challenge 9. The density is about 90.7%. The computation is

$$\frac{(\frac{1}{2})\pi r^2}{\sqrt{3}r^2} = \frac{\pi r^2}{2\sqrt{3}r^2} = \frac{\pi}{2\sqrt{3}}.$$

$\frac{\pi}{2\sqrt{3}}$ is approximately $.907 = 90.7\%$.

Packing Project 6

In the earlier projects, you observed that the triangle packing and the repeat triangle packing were engaged in a fierce rivalry to pack the most pennies. Some very interesting mathematical analysis will reveal more about this.

The triangle packing appeared to lose ground to the repeat triangle packing when the board was enlarged vertically. However, recall the growing space at the top of the board in the triangle packing at each step. It is important to figure out when there will be enough room to squeeze in an extra row of pennies. This is the first challenge of the next project. It requires the most effort. The remaining parts are mostly fun.

Challenge 10. Find out how many $6'' \times 6''$ squares must be placed vertically in a column for the triangle packing to squeeze in an extra row of pennies. In other words, how many boards are needed vertically to get the number of rows of pennies equal to a multiple of 9 plus 1?

Think about Challenge 10 for a while. Then start out on your own, or read on if you would like some ideas to help plan your solution.

To begin, consider the first $6'' \times 6''$ square. The height of the first row of pennies is $\frac{3}{4}''$. What is the height of the first two rows of pennies, the first three rows, etc.? Aha! We don't know! We need to compute how much height is added when a new row of pennies is added in the triangle packing.

Step-by-Step Outline for a Solution to Challenge 10

Step 1. Computation of the height of each new row in the triangle packing. Look at the figure of the basic equilateral triangle formed using the centers of three pennies. To help with the solution, we have enlarged the size of pennies in the diagram and added some line segments. Here is what we have added to the figure:

(i) the line segment AB, which is the altitude of the triangle extended to meet the top of the top circle;

(ii) EF a line segment tangent to the bottom two circles; and

(iii) EZ and FY, radii of the bottom circles to the points of contact at E and F with the tangent line.

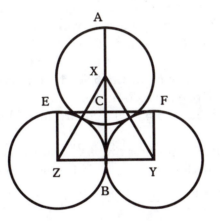

Notice that the added height you are trying to compute is exactly the length of the line segment AC. Can you find the length of AC? If you would like a few hints, read on.

Hints: (a) AC is the sum of the radius AX and the line segment XC. Since you know the radius has length $\frac{3}{8}''$, you need to find the length of XC.

(b) Remember that the basic triangle XYZ is equilateral. What is the length of its altitude XB?

(c) Now, if you can show that CB has length equal to $\frac{3}{8}''$, the length of the radius, you can compute the length of XC by subtraction. You may find it useful to use the fact that a tangent to a circle meets a radius at right angles.

Step 2. Computation of the height of the first row plus k additional rows of pennies.

Try this computation on your own.

Step 3. Put Steps 1 and 2 together to answer: How many $6'' \times 6''$ squares stacked vertically in a column are needed to squeeze in one extra row of pennies?

Try it on your own, or continue reading for more detailed suggestions.

The idea is to measure (and then estimate) the size of the space at the top as you increase the number of $6'' \times 6''$ boards. The answer to the question in Step 3 is the number of boards necessary for the space at the top to be at least equal to the height of one new row of pennies, which you computed in Step 1, and which we will estimate to be $.65''$.

Here is a start on the calculation. Remember that there are nine rows of pennies in the triangle packing of the $6'' \times 6''$ square. Let s_1 be the size of the space at the top of one triangle-packed board, s_2 the size of the space at the top of the triangle-packed $6'' \times 12''$ board made by stacking two square boards vertically, etc.

You know the height of the first row is $\frac{3}{4}''$ and the height of one additional row is $\sqrt{3}(\frac{3}{8})$, or approximately $.65$. The height of the space at the top of the first $6'' \times 6''$ board is

$$s_1 = 6 - \frac{3}{4} - 8\sqrt{3}\left(\frac{3}{8}\right).$$

Consequently, s_1 is at least $0.05''$.

Now continue the computation as you increase the number of triangle-packed boards stacked vertically and stop when the size of the space is at least $.65''$.

Excellent work, so far. The remaining two challenges of Project 6 are fun and reward all your hard work in the first part.

Challenge 11. Place 5 boards as in Challenge 10. Find out which packing, the triangle packing or the repeat triangle packing, wins the competition to pack the most pennies.

Challenge 12. Return to the SportsSmart $60'' \times 18''$ display shelf. Think of the shelf as composed of thirty $6'' \times 6''$ squares arranged so that there are 3 columns of 10 squares each with each column running from one end of the $60''$ shelf to the other end.

Using either the triangle packing or the repeat triangle packing, compute the number of pennies packed on the shelf if you start at one end as indicated.

Solutions

Step 1. Computation of the height of each new row in the triangle packing. The length of AC is $\sqrt{3}(\frac{3}{8})''$, or approximately $.65''$. Here is one method of calculating this: From the fact that tangents are perpendicular to radii at the point of contact, it follows that $EFYZ$ is a rectangle. Since the altitude XB meets YZ at right angles, the length of CB is equal to that of the radii EZ and FY, namely $\frac{3}{8}$. Now the altitude XB has length $\sqrt{3}(\frac{3}{8})$, so XC has length $\sqrt{3}(\frac{3}{8}) - (\frac{3}{8})$, and finally, the length of AC is $\sqrt{3}(\frac{3}{8}) - (\frac{3}{8}) + (\frac{3}{8}) = \sqrt{3}(\frac{3}{8})''$.

Step 2. Computation of the height of the first row plus k additional rows of pennies. The height of the first row of pennies is $\frac{3}{4}''$ so using the previous computation, the height of the first row plus one additional row is $\frac{3}{4}'' + \sqrt{3}(\frac{3}{8})''$. The height of the first row plus two additional rows is $\frac{3}{4}'' + \sqrt{3}(\frac{3}{8})'' + \sqrt{3}(\frac{3}{8})''$ and the height of the first row plus k additional rows is $\frac{3}{4}'' + k\sqrt{3}(\frac{3}{8})''$, or approximately $.75'' + (.65)k''$.

Step 3. Continuation of the computation of the size of the space at the top of the vertically stacked, triangle packed boards.

Place a second square on top of the first one and add nine new rows of pennies in the triangle packing. The space at the top of this new board is

$$s_2 = 12 - \frac{3}{4} - 8\sqrt{3}(\frac{3}{8}) - 9\sqrt{3}(\frac{3}{8}) = 12 - \frac{3}{4} - 17\sqrt{3}(\frac{3}{8}),$$

so s_2 is at least $.20''$, but still much smaller than $.65''$. There is not enough room for another row of pennies.

Add a third square, pack it with nine more rows, and calculate the space at the top. It is

$$s_3 = 18 - \frac{3}{4} - 26\sqrt{3}\left(\frac{3}{8}\right),$$

so s_3 is at least $.35''$ but still not big enough.

Add a fourth square, pack it with nine more rows. This time the space at the top is

$$s_4 = 24 - \frac{3}{4} - 35\sqrt{3}\left(\frac{3}{8}\right),$$

so s_4 is at least $.50''$. One more board ought to do it. Let's see.

The addition of nine new rows of pennies on the fifth square leaves a space at the top of

$$s_5 = 30 - \frac{3}{4} - 44\sqrt{3}\left(\frac{3}{8}\right),$$

so s_5 is at least $.65''$, the number needed for an extra row.

Challenge 11. The triangle packing wins! On a $6'' \times 30''$ board composed of five vertically stacked $6'' \times 6''$ squares using the repeat triangle packing, $5 \times 68 = 340$ pennies are packed.

On the $6'' \times 30''$ board, using the triangle packing, $68 + 67 + 68 + 67 + 68 + 7 = 345$ pennies are packed.

Challenge 12. Think of the $60''$ long and $18''$ wide shelf as composed of thirty $6'' \times 6''$ squares arranged so that there are 3 columns of 10 squares each.
Triangle packing: Each set of five squares stacked vertically allows an extra row of pennies, so there are at least $(9 \times 5) + 1 + (9 \times 5) + 1 = 92$ rows of pennies. Is there possibly room for one more row? To check, compute the height of the 92 rows of pennies and subtract that number from $60''$, the height of the shelf. The height of 92 rows of pennies is: $\frac{3}{4}'' + 91\sqrt{3}\left(\frac{3}{8}\right)''$, or approximately $59.86''$. But $60'' - 59.86'' = .14''$ is less than the height required for an additional row of pennies. This means that there are exactly 92 rows of pennies.

In each of the three columns (of 10 squares) 46 of the rows are rows of 8 pennies and 46 are rows of 7 pennies. Now you will be able to fit in extra pennies at each column juncture: one extra penny for each row of 7 pennies. So at each column juncture, you can fit in 46 extra pennies. Thus the total number of pennies packed is: $(3 \times 690) + (2 \times 46) = 2162$. (The density of this packing is about 88.4%.) Compare with the packing you did in Project 4 where you packed the shelf from the back to the front.
Repeat triangle packing: In each column of squares you can pack $10 \times 68 = 680$ pennies. There are $10 \times 5 = 50$ rows of 8 pennies and $10 \times 4 = 40$ rows of 7 pennies. You will be able to fit in extra pennies at each column juncture: one extra penny for each row of 7 pennies. So at each column juncture, you can fit in 40 extra pennies. Thus the total number of pennies packed is $(3 \times 680) + (2 \times 40) = 2120$. (The density of this packing is about 86.7%.) The triangle packing is a big winner here!

Suggestions for the Endurance Athlete

10K Challenge. Investigate the following variation of the triangle packing.

   In the definition of the repeat triangle packing on a vertical stack of $6'' \times 6''$ boards, the triangle packing repetitions start at the bottom of each added square. This means that there is a space after every nine rows of this packing. A tighter packing, called the *tight repeat triangle packing* can be defined by eliminating these spaces. For this packing, after every nine rows, the next row is laid directly on top of the previous row as follows:

   You see that in the tight repeat triangle packing after a certain number of boards are added vertically in a column, there will be enough room for an extra row of pennies. How many boards are needed to squeeze in an extra row of pennies? This is one of many questions you can ask about this packing. Investigate the tight repeat triangle packing and make comparisons with the triangle and repeat triangle packings.

20K Challenge. Recall that the density of the triangle packing on the whole plane is approximately 90.7%. You observed that the density tends to get closer and closer to 90.7% as larger and larger areas of the plane are packed. Here is an excellent question to investigate. What is the smallest rectangular area which, when covered with pennies in the triangle packing, gives a density of 90% or better? Then try for a density of 90.5% or better.

30K Challenge. Let n be any positive integer and let S be a 1 unit $\times$ 1 unit square. Find a packing of identical non-intersecting disks onto the square so that the sum of their radii is greater than n units.

40K Challenge. For $n = 1, 2, \ldots, 16$, find the side length s, not necessarily an integer, of the smallest square S such that n disks of radius 1 unit can be packed onto S.

50K Challenge. Think again about those tennis balls at SportSmart. Suppose you remove the balls from their cylindrical cans and pack the balls in a square box. What arrangements of balls gives the greatest density of balls in the box? (The problem of finding a sphere packing of the greatest density is an old and very difficult problem that has just recently been solved.)

References

More information about the 40K Challenge can be found on the web at Erich's Packing Center: www.stetson.edu/~efriedma/packing.html.

For more information on the 50K Challenge, see Gardner, M., *The Colossal Book of Mathematics*, Chapter 10: Packing Spheres. Two web sites with more information on this topic are mathworld.wolfram.com/KeplerConjecture.html as well as www.stetson.edu/ edfriedma/packing.html.

The focus in this section has been packing circles into squares. For a discussion of packing circles in circles, circles in triangles, squares in triangles, etc., see www.maa.org/mathland/mathland_11_25.html, for example, as well as www.stetson.edu/~edfriedma/packing.html.

Chapter Nine

Lattice Polygons

The arrangement of pegs on a geoboard is an example of a lattice. The starting point of this activity is the construction of geometric figures on a geoboard or a lattice, and an investigation of their properties, such as perimeter and area. This leads to the discovery that there are certain familiar geometric figures that cannot be constructed on a lattice.

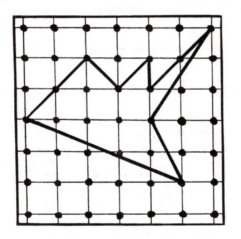

"Poincaré, on being asked how he made discoveries, answered 'by thinking about them often.' "

J. Dieudonné

Have you ever played the game of dots, or made figures by stretching elastic bands around the pegs on a geoboard? In The Game of Dots, players start with dots evenly spaced on a piece of paper, and connect two dots at a time, horizontally or vertically, to form the sides of squares.

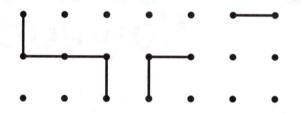

The arrangement of dots in the game and pegs on a geoboard are examples of lattices. You will discover that many, but not all, geometric figures can be constructed on dotted paper or on a geoboard, making it easy to investigate properties such as perimeter and area.

In mathematical terms, a rectangular array of dots or points spaced so that the horizontal and vertical distance between dots is the same is called a *lattice*. The dots or points are called *lattice points*. Polygons whose vertices are lattice points are called *lattice polygons*. Some lattice polygons have lattice points other than vertices on their sides, others do not.

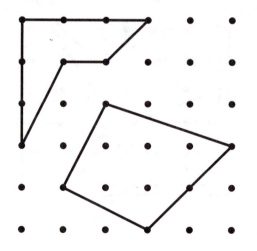

In this investigation of lattice polygons, you will investigate which polygons can be lattice polygons. You will also discover a neat way to compute the area of a lattice polygon.

Part I: Lattice Polygon Warm-Up

Some supplies are needed. Put a square array of, say, 10 rows of 10 equally spaced dots on each of six sheets of paper.

We will begin with some experimentation.

Challenge 1. Can you construct a lattice square? In other words, can you construct a square with each vertex at a dot on your lattice?

How about a lattice rectangle that is not a square?

Can you construct a lattice triangle?

How about a lattice right triangle?

How about a lattice isosceles triangle?

Can you construct a lattice equilateral triangle?

Was it smooth going until you hit the equilateral triangle? Later, you will see why the equilateral triangle causes difficulty. So far, you have a good start; you know that some squares, rectangles, triangles, right triangles, and isosceles triangles are lattice polygons.

Challenge 2. Try to construct a lattice kite, a lattice house pentagon, and a lattice hexagon.

Note that there are no rules that require that any of the sides of your polygons have to be horizontal or vertical, even if the sides meet at right angles. Let's call a lattice square *skew* if no side is either horizontal or vertical.

Challenge 3. Try to construct a skew square.

Why does the construction of a lattice equilateral triangle by connecting dots cause such a problem? The reason has to do with its area. Digging into this further involves computing areas of some polygons. Here is a short Area Warm-Up.

Area Warm-Up

For this review, there is just one basic area formula to remember; all the others will follow from it.

The *area A* of a rectangle is the *product* of the lengths of its *base b* and its *altitude h*: $A = bh$.

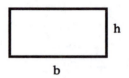

Since a square is a special rectangle with base and altitude of equal length, the formula for the area of a square with side of length s is often written $A = s^2$.

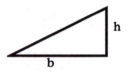

Challenge 4. What comes next? Triangles. First, do you know how to use the formula for the area of a rectangle to compute the area of the right triangle pictured above with base of length b and altitude of length h?

The area of a right triangle is one half the product of the lengths of its base and its altitude: $A = \frac{1}{2}bh$.

You may know that the area of *every* triangle with base of length b and altitude of length h is $A = \frac{1}{2}bh$. How do you show this is true?

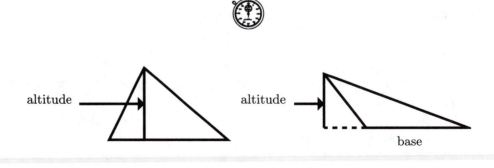

Did you remember that the altitude of a triangle to a given base is the perpendicular line segment to the line of the base from the vertex not on the base?

If the altitude is one of the sides of the triangle, then the triangle is a right triangle and you know what its area is.

Challenge 5. If the foot of the altitude falls inside the triangle, then what is the triangle's area? If the foot of the altitude falls outside the triangle, then what is the triangle's area?

If you were able to figure this out, then you have completed the proof that the *area* of a triangle is *one-half* the *product* of the lengths of its *base b* and *altitude h*: $A = \frac{1}{2}bh$.

Now you can find the area of lots of polygons. Take a minute to recall the definition of a polygon: Start with n distinct points $P_1, P_2, \ldots, P_n$ in a plane, where n is at least three, so that no three consecutive points lie on a line. If the segments $P_1P_2, P_2P_3, \ldots, P_{n-1}P_n, P_nP_1$ intersect only at their endpoints, they form a polygon.

Remember that a polygon is convex if the line segment joining any pair of vertices lies within the polygon. If the polygon is convex, we can break up the polygon into non-overlapping triangles. Start at any vertex, say P_1, and join P_1 to $P_3, P_4, \ldots, P_{n-2}, P_{n-1}$, that is, to all the other vertices except the two on either side of P_1. Since the polygon is convex, all of these line segments lie within the polygon.For example, a pentagon may be broken up this way:

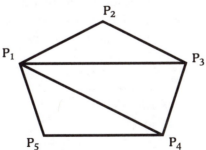

In mathematical terms, the polygon has been *triangulated.* A triangulation of a polygon is a cutting up (or dissection) of the polygon into triangles *with vertices that are those of the original polygon.* Here is another triangulation of the pentagon:

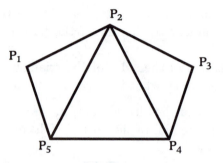

The area of the polygon is the sum of the areas of the triangles in a triangulation. So, if possible, you want to triangulate the polygon in a way that produces "good" triangles, that is, triangles whose lengths of bases and altitudes you know or can compute. Try your hand at triangulating and computing area.

Challenge 6. Show that the area A of a parallelogram with base of length b and altitude of length h is $A = bh$ by triangulating the parallelogram.

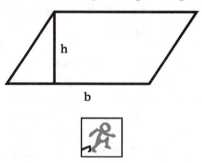

Challenge 7. Do you remember that the formula for the area of a trapezoid is one half the sum of the lengths, b_1 and b_2, of its bases times the length h of its altitude? Show that you can derive this formula by triangulating the trapezoid.

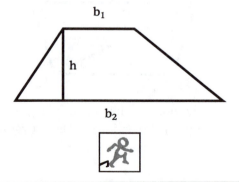

It is true that every polygon can be triangulated, but of course, it is trickier to triangulate a non-convex polygon. Draw your favorite nasty-gon, such as this one,

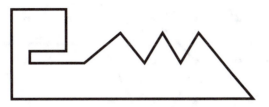

Can you triangulate it? Can you compute the area of your nasty-gon?

Did you see that its area is the sum of all the areas of the triangles in your triangulation? You can compute the area *if*, and sometimes this is a big *if*, you can compute the areas of all the triangles.

The nasty-gon can be triangulated like this, for example:

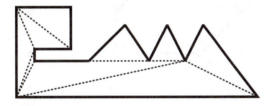

Its area is the sum of nine triangles and can be computed *if* the lengths of all nine bases and altitudes can be calculated.

This completes the Area Warm-Up.

Solutions

Challenge 1. Here are some examples of lattice polygons:

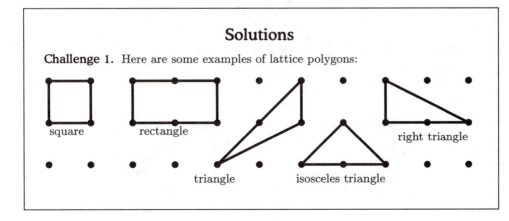

Challenge 2. Here are examples of a lattice kite, a lattice house pentagon, and a lattice hexagon:

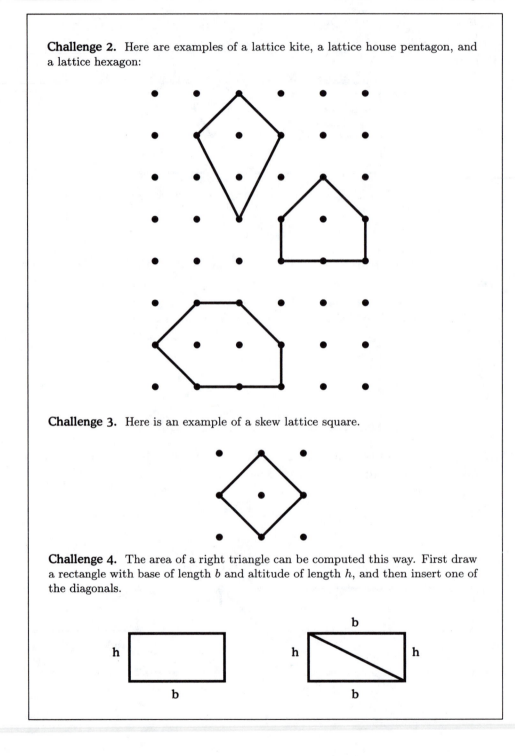

Challenge 3. Here is an example of a skew lattice square.

Challenge 4. The area of a right triangle can be computed this way. First draw a rectangle with base of length b and altitude of length h, and then insert one of the diagonals.

This decomposes the rectangle into two right triangles, each having base of length b and altitude of length h. These right triangles are congruent because corresponding sides are equal. This means that the areas of the right triangles are equal, and the area of the rectangle is two times the area of the right triangle. In other words, the area of the right triangle with base of length b and altitude of length h is one half the area of the rectangle. The area of a right triangle is then one half the product of the lengths of its base and its altitude: $A = \frac{1}{2}bh$.

Challenge 5. If the foot of the altitude falls inside the triangle, then the area of the triangle is the sum of the areas of two right triangles whose bases have no overlap and taken together form the base of the original triangle. The length of the base b of the triangle is the sum of the lengths of the bases b_1, b_2 of the right triangles.

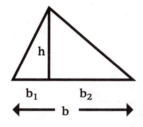

So, using the formula of the area of a right triangle, the area A of the original triangle is

$$A = \frac{1}{2}b_1 h + \frac{1}{2}b_2 h = \frac{1}{2}(b_1 + b_2)h = \frac{1}{2}bh.$$

If the foot of the altitude falls outside the triangle, the original triangle becomes part of a big right triangle. The area of the original triangle added to the area of the small right triangle is the area of the big right triangle.

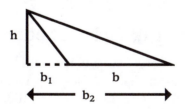

So the area of the original triangle is found by subtracting the area of the small right triangle from that of the big right triangle. The area is

$$A = \frac{1}{2}b_2 h - \frac{1}{2}b_1 h = \frac{1}{2}(b_2 - b_1)h = \frac{1}{2}bh,$$

since the length of the base b of the original triangle is the difference $b_2 - b_1$ of the lengths of the bases of the right triangles.

Challenge 6. To check the formula for the area of a parallelogram with base of length b and altitude of length h, cut up the parallelogram in the following way:

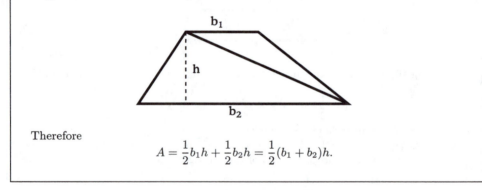

The parallelogram is cut up (or dissected) into two triangles having base b and height h. Therefore

$$A = \frac{1}{2}bh + \frac{1}{2}bh = bh.$$

Challenge 7. The trapezoid with bases b_1 and b_2 and height h can be dissected into two triangles. One has base b_1 and height h and the other has base b_2 and height h.

Therefore

$$A = \frac{1}{2}b_1h + \frac{1}{2}b_2h = \frac{1}{2}(b_1 + b_2)h.$$

Part II: Pick's Theorem

As we have observed, the area of any polygon can be calculated if the lengths of the bases and altitudes of the triangles in a triangulation of the polygon are known. However, this data may be very difficult to obtain. Lattice polygons are special. There is a formula for computing the area of any lattice polygon that simply requires counting lattice points. It is called Pick's theorem.

We now embark on the discovery of Pick's theorem. The formula expresses the area of a lattice polygon in terms of the number of lattice points inside the polygon and the number of lattice points on its sides. We use the word *boundary* for the collection of all of the sides (or edges) of the polygon. This will allow us to distinguish between those lattice points on the boundary and those lattice points inside (or in the interior of) the polygon.

Let's experiment first with squares. The horizontal or vertical distance between the evenly spaced dots on your lattices will be the unit of measurement. We will call it 1 unit.

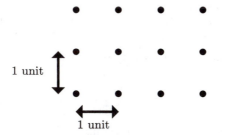

Challenge 8. On one of your lattices, make a 3 unit × 3 unit square with horizontal and vertical sides. What is its area?

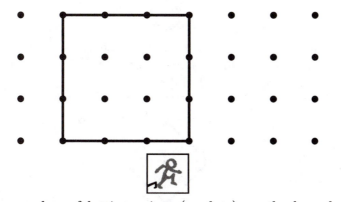

Look at the number of lattice points (or dots) on the boundary of your square and the number inside your square. Find an equation relating the area $A = 9$ to the number of lattice points on the boundary and the number of lattice points inside.

Let B be the number of lattice points on the boundary of the square and I the number inside your square. Can you write 9 in terms of B and I? Can you guess what the area of any lattice square is in terms of B and I?

Now put your guess to the test. Make a 2 unit × 2 unit square (with horizontal and vertical sides) on your lattice. You know the area is 4 square units. Does your formula prove correct here?

More testing is necessary. Draw a 5 unit × 5 unit square (with horizontal and vertical sides). Study all three squares. Find B and I for each square, and see if you can find some new ideas for connecting the area A to B and I.

Does your formula check out for a 1 unit × 1 unit square (with horizontal and vertical sides)?

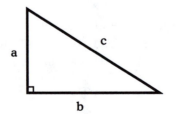

Test your formula on this skew square. Each side of this square is the diagonal of a 1 unit × 1 unit lattice square with horizontal and vertical sides. What is the length of such a diagonal?

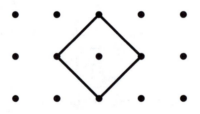

Hint: You may want to use the Pythagorean theorem. Remember that it states: In a right triangle, the square of the length of the hypotenuse is equal to the sum of the squares of the lengths of the legs.

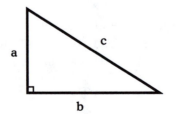

So if the right triangle has hypotenuse of length c and legs of lengths a and b, then $c^2 = a^2 + b^2$.

Did You Find Pick's Formula?

Remember that B is the number of lattice points on the boundary of the square and I is the number of lattice points inside. Your formula for the area A may be in a different form but it should correspond to the following: $A = \frac{B}{2} + I - 1$.

Let's verify that the above formula is the correct one for all the lattice squares in the experiment.

For the 3 unit × 3 unit square, the area is $A = 9$ square units. For the 3 unit × 3 unit square, $B = 12$ and $I = 4$, so $\frac{B}{2} + I - 1 = 6 + 4 - 1 = 9$ is the correct area A.

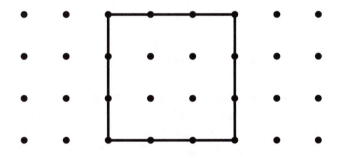

For the 2 unit × 2 unit square, $B = 8$ and $I = 1$, so $\frac{B}{2} + I - 1 = 4$ is the correct area A.

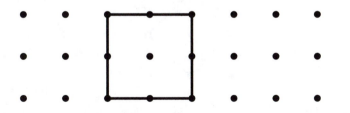

For the 5 unit × 5 unit square, the area $A = 25$ square units. Since $B = 20$, and $I = 16$, and $\frac{B}{2} + I - 1 = 10 + 16 - 1 = 25$, the formula is correct.

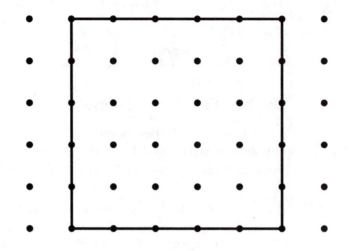

For the 1 unit × 1 unit square, the area $A = 1$ square unit. Since $B = 4$ and $I = 0$, and $\frac{B}{2} + I - 1 = 2 + 0 - 1 = 1$, the formula is correct.

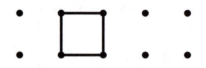

The skew square has each side equal to the diagonal of a 1 unit × 1 unit square. By the Pythagorean theorem, the length s of each side is $\sqrt{2}$ units. The skew square is a $\sqrt{2} \times \sqrt{2}$ square and has area $A = (\sqrt{2})^2 = 2$ square units. For the skew square, $B = 4$, $I = 1$ and $\frac{B}{2} + I - 1 = 2 + 1 - 1 = 2$, which is the area. The formula checks.

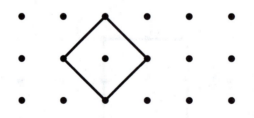

What about the area of lattice polygons other than the squares?

Do you think the formula $A = \frac{B}{2} + I - 1$ will hold for the following trapezoid? As before, B is the number of lattice points on the boundary and I the number of lattice points inside.

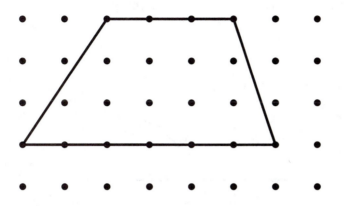

Challenge 9. Check to see if the formula works for this lattice trapezoid. Notice that the slanted sides of the trapezoid don't have many lattice points on them.

It can be proved that Pick's formula is true for every lattice polygon.

Pick's Theorem. *The area A of a lattice polygon is one less than the sum of half the number B of lattice points on the boundary and the number I of lattice points inside:*

$$A = \frac{B}{2} + I - 1.$$

Here is a good game based on Pick's theorem.

Pick's Polygon Game

Pick's Polygon Game is a game for any number of players.

Directions. The players take turns being Pick. Whoever is Pick draws a lattice polygon and challenges the other players to construct a different lattice polygon *having the same area*. The first player to find a polygon that answers the challenge correctly within five minutes earns a number of points equal to the area of the polygon. If no player is able to find such a polygon within the time limit and if Pick can do it, then Pick wins the points.

A harder version of the game allows the player being Pick to have the option of making the challenge more difficult by requiring the polygons to have *certain other properties*, such as a specified shape or a specified number of boundary points, *in addition to having the same area*.

I'll be Pick to start. Here is a lattice polygon:

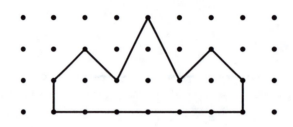

Challenge 10. Find another lattice polygon with the same area.

Did you find another lattice polygon of area 10? If so, you earn 10 points. Some polygons with area 10 square units are found in the Solutions. Here are two more:

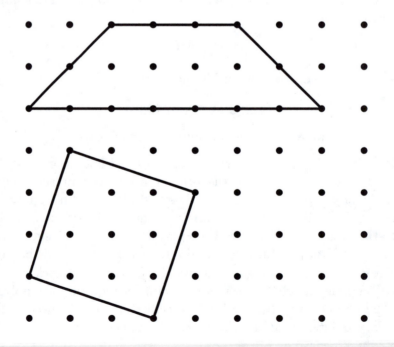

Challenge 11. The skew figure in the previous illustration looks like a square. Is it? Using Pick's formula, you can see it has area 10 square units. Is each side of length $\sqrt{10}$ units?

Did you notice that each side of the skew square is the hypotenuse of a right triangle with one leg of length _____ and the other leg of length _____?

Challenge 12. Now, let's try the harder version of the game. I'll be Pick, and I am going to add a specified shape to my challenge. Find a lattice triangle that has the same area as the lattice polygon below:

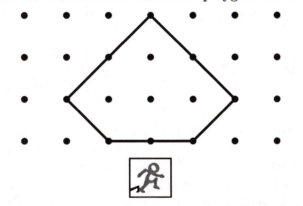

Challenge 13. I'll take one more turn being Pick. Find a lattice polygon with exactly 6 lattice points on its boundary that has the same area as the lattice "X" below.

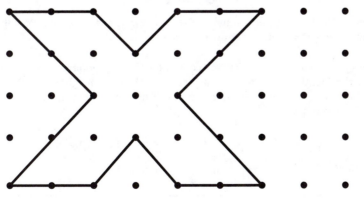

Did you find a lattice polygon with the same area that has exactly six lattice points on its boundary? If so, since the "X" has area 14 square units, you earn 14 points.

Now it's your turn to play Pick. If no one else is around, you can play solo. Draw a lattice polygon and challenge yourself to find others having the same area, and, perhaps, some other properties too. Alphabet lattice polygons, like the "X" above, are fun examples.

We conclude Part II by returning to the problem of constructing a lattice equilateral triangle. Why did we have such difficulty? Here is the reason.

It is impossible to construct a lattice equilateral triangle.

Why? One explanation is that the area of an equilateral triangle is the wrong type of number. It is a type of number that can never satisfy Pick's theorem. For more about this, see the Heavy Lifting section at the end of the chapter.

If you are really, really, curious about the equilateral triangle right now, pause for a while and ask yourself these questions. What kinds of numbers can be areas of lattice polygons? Pick's theorem says they are of the form $\frac{B}{2} + I - 1$, so what do such numbers look like? Suppose that an equilateral triangle can be constructed in a lattice. Can its area be of this form?

Solutions

Challenge 8. The area of a square is s^2, where s is the length of its side. Since this square has $s = 3$ units, its area is 9 square units.

Challenge 9. The area of the lattice trapezoid is $\frac{27}{2}$ square units. Using the sum of the lengths of its bases times the length of its height, you obtain $A = \frac{1}{2}(3 + 6)(3) = \frac{27}{2}$. So $A = \frac{27}{2}$ square units. Since the trapezoid has $B = 11$ and $I = 9$, it follows that

$$\frac{B}{2} + I - 1 = \frac{1}{2}(11) + 9 - 1 = \frac{11}{2} + 8 = \frac{11}{2} + \frac{16}{2} = \frac{27}{2}.$$

The formula gives the correct number for the area.

Challenge 10. The polygon

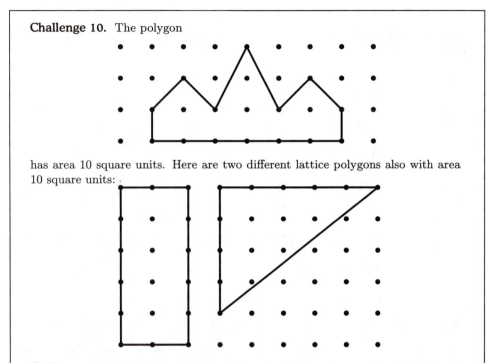

has area 10 square units. Here are two different lattice polygons also with area 10 square units:

Challenge 11. Each side of the skew figure is the hypotenuse of a right triangle with legs of length 1 unit and 3 units. So each side has length $\sqrt{10}$ units.

Challenge 12. This time Pick's polygon has area 7 square units. Did you find a lattice triangle with area 7? If so, you earn 7 points. Here is an example of a lattice triangle that works:

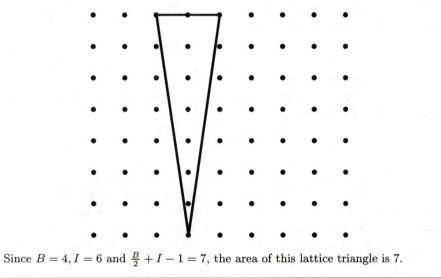

Since $B = 4, I = 6$ and $\frac{B}{2} + I - 1 = 7$, the area of this lattice triangle is 7.

Challenge 13. Here's a lattice kite that has area 14 square units and exactly 6 lattice boundary points.

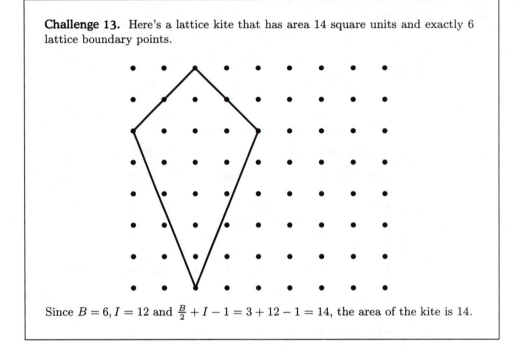

Since $B = 6, I = 12$ and $\frac{B}{2} + I - 1 = 3 + 12 - 1 = 14$, the area of the kite is 14.

Part III: What Numbers Are Areas of Lattice Squares?

In the next part of this investigation, we will explore an intriguing question. *What numbers can occur as areas of lattice squares?*

We begin with an experiment.

Investigate which of the integers

$$1, 2, 3, 4, 5, 6, 7, 8, 9, 10$$

are areas of lattice squares.

Hint: Seven of the integers are areas of lattice squares and three cannot be areas of lattice squares.

Did you see that $1, 4$, and 9 are areas of lattice squares? Good. These numbers are examples of what are called perfect squares: $1 = 1^2, 4 = 2^2$ and $9 = 3^2$. A *perfect square* is a positive integer that is the square of another integer. By definition, then, every perfect square N has the form $N = s^2$ for some positive integer s. So N is the area of an s unit $\times$ s unit lattice square with horizontal and vertical sides.

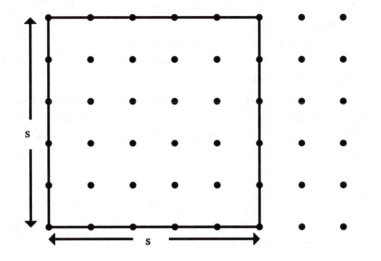

It follows that

- 1 is the area of a 1 unit × 1 unit lattice square (with horizontal and vertical sides);

- 4 is the area of a 2 unit × 2 unit lattice square (with horizontal and vertical sides); and

- 9 is the area of a 3 unit × 3 unit lattice square (with horizontal and vertical sides).

How about 2? Did you remember that 2 is the area of the skew square encountered earlier?

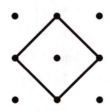

Each side has length $\sqrt{2}$ units, the length of a diagonal of a 1 unit × 1 unit square, and its area is $(\sqrt{2})^2 = 2$ square units.

The numbers $1, 2, 4, 5, 8, 9,$ and 10 are areas of lattice squares, but the numbers $3, 6,$ and 7 are not. Lattice squares with areas 5 square units, 8 square units, and 10 square units are more complicated to construct than lattice squares with area equal to a perfect square. We will learn how to construct these lattice squares.

As the investigation proceeds, you will discover which positive integers can be areas of lattice squares. Even better, you will find that each of these numbers contains the blueprint for the construction.

Since constructing lattice squares with area equal to a perfect square is easy, the investigation will focus on the numbers that are not perfect squares, but are areas of lattice squares. What do these numbers look like? What properties do they have?

To find out, let n be a positive integer and assume that

1. n is not a perfect square;

2. there is a lattice square of area n square units.

Constructing a lattice square of area n square units is the same as constructing a lattice square with side of length $\sqrt{n}$ units. If n is not a perfect square, $\sqrt{n}$ is not an integer. A lattice square of area n square units will usually look different from one with area that is a perfect square. Let's see why.

Question 1. Suppose a line segment connects two lattice points, and it has length that is not an integer. For example, it could have length $\sqrt{2}$ or $\sqrt{5}$. How must such a line segment be positioned in the lattice? Horizontal, vertical, or skew? (We will call a line segment with lattice points as endpoints *skew* if it is neither horizontal or vertical.)

Hint: Think about what kinds of numbers give the horizontal and vertical distances between dots.

Question 2. What does this mean for lattice squares? If a lattice square has area n square units and n is not a perfect square, how will the square be positioned in the lattice?

The final question reveals the essential characteristic of n. Suppose that n is not a perfect square so $\sqrt{n}$ is not an integer, and suppose $\sqrt{n}$ units is the length of a skew side of a lattice square.

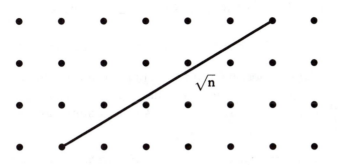

Keep the Pythagorean theorem in mind and try to answer the following question.

Question 3. What property does n have that is related to squares of integers?

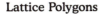

The answer to the above question merits repeating. If n is not a perfect square, $\sqrt{n}$ units can be the length of a side of a lattice square only if n can be written as a sum of squares of two positive integers: $n = a^2 + b^2$. In fact, as you observed, $\sqrt{n}$ is the length of the hypotenuse of a right triangle where a is the length of the horizontal leg and b is the length of the vertical leg.

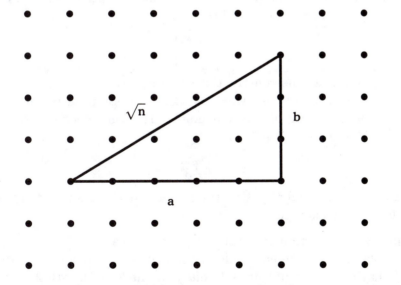

Note that a perfect square s^2 can be written $s^2 + 0^2$. This means that the restriction "if n is not a perfect square" may be eliminated if we allow one of the squares to be 0. The following statement summarizes the investigation so far:

> If n is a positive integer, n (square units) can be the area of a lattice square only if n can be written as a sum of squares of two non-negative integers.

Question 4. Earlier, by experimentation, you observed that lattice squares of areas $1, 2, 4, 5, 8, 9,$ and 10 (square units) can be constructed. Check that each of these numbers is a sum of squares of two non-negative integers.

Question 5. Which of the integers between 11 and 20 are sums of squares of two non-negative integers?

Question 6. Can you find a positive integer that is both a perfect square and a sum of squares of two positive integers?

Next, we investigate the construction of these lattice squares. Suppose the positive integer $n = a^2 + b^2$, where a and b are non-negative integers; can you construct a square of side $\sqrt{n}$ units? Try some special cases such as $10 = 3^2 + 1^2$ and $13 = 3^2 + 2^2$ first.

Here is a neat method for constructing the squares that is easy to remember. First, let's do an example.

Example 1. Construction of a lattice square of area $13 = 3^2 + 2^2$.

The plan is to make each side of the square the hypotenuse of a right triangle having one leg of length 3 units and one leg of length 2 units. Each side would then have length equal to $\sqrt{3^2 + 2^2} = \sqrt{13}$ units, and the square would have area equal to $(\sqrt{13})^2 = 13$ square units.

Construction.

1. Add 3 and 2 to get $3 + 2 = 5$.

2. Draw a 5 unit $\times$ 5 unit lattice square with horizontal and vertical sides.

3. Start at the upper left hand corner, move clockwise around the square and mark the fourth dot, or three units, along each side as in the picture below.

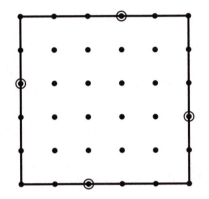

4. Connect the marked dots on the sides.

You have constructed a lattice square of area 13 sq. units, since each side has length equal to $\sqrt{3^2 + 2^2} = \sqrt{13}$ units.

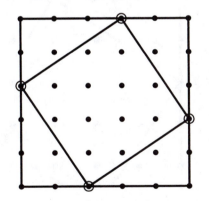

Isn't that neat? The clue to construction is coded in the display: $13 = 3^2 + 2^2$.

The general case is just as easy.

General Case. Construction of a lattice square of area $n = a^2 + b^2$.

Suppose $n = a^2 + b^2$ with a and b non-negative integers. If $a = 0$ or $b = 0$, then n is a perfect square, and we have already discussed how to construct a lattice square of area n. So we will assume that both a and b are positive integers. To construct a lattice square of area n square units, the plan is to construct a lattice square with each side equal to the hypotenuse of a right triangle with one leg of length a units and the other leg of length b units. Then each side will have length $\sqrt{n} = \sqrt{a^2 + b^2}$.

Construction.

1. Add a and b.

2. Draw an $(a+b)$ unit $\times$ $(a+b)$ unit lattice square with horizontal and vertical sides.

3. Start at the upper left hand corner, move clockwise around the square and mark the $(a+1)$st dot, or a units, along each side as in the picture below.

4. Connect the marked dots on the sides.

A lattice square with sides of length $\sqrt{n}$ units is constructed. Each side is the hypotenuse of a right triangle with legs of length a and b.

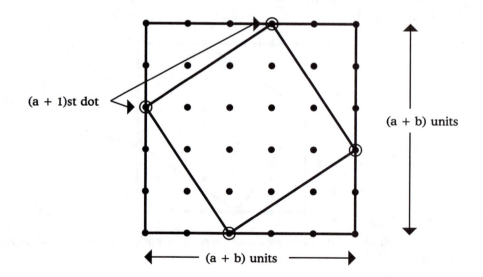

(a + 1)st dot

(a + b) units

(a + b) units

Your work constitutes a proof of the following theorem:

Theorem. *Let n be a positive integer. If n (square units) is the area of a lattice square, then n is the sum of squares of two non-negative integers. Conversely, if n is the sum of squares of two non-negative integers, then n (square units) is the area of a lattice square.*

Solutions

Question 1. Since all horizontal and vertical distances between dots in the lattice are integers, a line segment of non-integer length will be skew.

Question 2. If a lattice square has area n square units and n is not a perfect square, each side of the square will be skew, that is, neither horizontal nor vertical. It will be a skew square.

Question 3. If n is not a perfect square and n is the area of a lattice square, then n has the following property related to squares of integers. Each skew side of the square is the hypotenuse of a right triangle with one horizontal leg and one vertical leg. This means that n is the sum of two squares of positive integers $n = a^2 + b^2$, where a is the length of the horizontal leg and b is the length of the vertical leg.

Question 4. Here are the integers $1, 2, 4, 5, 8, 9, 10$ written as sums of two squares of non-negative integers: $1 = 1^2 + 0^2, 2 = 1^2 + 1^2, 4 = 2^2 + 0^2, 5 = 2^2 + 1^2, 8 = 2^2 + 2^2, 9 = 3^2 + 0^2$ and $10 = 3^2 + 1^2$.

Question 5. There are exactly five integers between 11 and 20 that are sums of two squares of non-negative integers: $13 = 3^2 + 2^2, 16 = 4^2 + 0^2, 17 = 4^2 + 1^2, 18 = 3^3 + 3^3$ and $20 = 4^2 + 2^2$.

Question 6. Here are some positive integers that are both perfect squares and sums of two squares of positive integers: $25 = 9 + 16$ and $100 = 36 + 64$ and $169 = 25 + 144$. This means that for each such number n, there is a lattice square with horizontal and vertical sides of area n and a skew lattice square of area n.

Part IV: Heavy Lifting

In this section you will examine examples of regular polygons that cannot be lattice polygons.

The Equilateral Triangle

The time has come to find out why it is impossible to construct a lattice equilateral triangle. We will consider two methods of proof. Both methods use proof by contradiction. In fact, both proofs assume that a lattice equilateral triangle can be constructed, and then arrive at the false statement that $\sqrt{3}$ is a fraction. You know that $\sqrt{3}$ is not a fraction. The techniques used in the proofs are, as you will see, quite dissimilar.

Theorem. *There does not exist a lattice equilateral triangle.*

Proof 1. This proof uses Pick's theorem. Recall that Pick's theorem states that the area A of a lattice polygon is given by the following formula

$$A = \frac{B}{2} + I - 1,$$

where B and I are the number of lattice points on the boundary and in the interior, respectively.

Step 1. Begin the proof by identifying exactly what types of numbers B, I and A are. Try to specify whether they are positive, negative, non-negative, etc., and whether they are integers, non-integers, fractions, non-fractions, real numbers, etc.

Step 2. Can you figure out something special about the length s of a side of any lattice polygon? You know that the length s might not be an integer, but what can you say about s^2? Why?

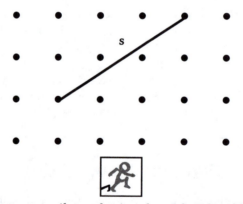

Step 3. Assume that an equilateral triangle with side of length s is a lattice triangle. By Step 1 and Step 2, you know that the area A has certain properties. Compute the area A using the standard formula. Try to derive a contradiction.

Solutions

Step 1. B is always a positive integer. I might be 0 but is always a non-negative integer. A is always a positive number but it might not be an integer. At best, A is an integer; at worst, A is a fraction with denominator 2, that is, one-half of an integer.

Step 2. The square of s, s^2, will always be a positive integer. For if a side of a lattice polygon is horizontal or vertical, its length s will be a positive integer and s^2 will also be a positive integer. If the side is skew, it will be the hypotenuse of a right triangle with legs that have integer lengths, say a and b. So $s^2 = a^2 + b^2$, and s^2 is an integer.

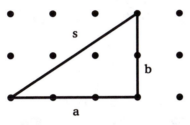

Step 3. Computing the area A in the standard way gives $A = \frac{\sqrt{3}}{4}s^2$, since the base of the triangle is s and the height is $\frac{\sqrt{3}}{2}s$. But, by Pick's Theorem, A is a fraction. So $\frac{\sqrt{3}}{4}s^2 = \frac{m}{n}$, where m and n are integers. Thus, $\sqrt{3} = \frac{4m}{ns^2}$, and, since $4, m, n$ and s^2 are all integers, you arrive at the false statement that $\sqrt{3}$ is a fraction. This contradiction shows that it is impossible to construct a lattice equilateral triangle.

Proof 2. This proof is for those who have had some experience using co-ordinates to locate points in the plane, and who know that the length of the line segment between two points (p, q) and (r, s) in the plane is $\sqrt{(r - p)^2 + (s - q)^2}$.

Imagine having a lattice that covers the entire plane. The points on the lattice are exactly the ones on the plane that have integers for both coordinates. As in the first proof, assume that a lattice equilateral triangle can be constructed. You will use coordinates to argue to the same false statement that $\sqrt{3}$ is a fraction.

Choose one of the vertices of the triangle as the origin $(0, 0)$. Call that vertex O. Name the other two vertices A and B and draw the altitude from A to OB, as shown below: (The pictured triangle doesn't look equilateral, does it? Of course, that's exactly the point. You are going to show that the pictured triangle can't be equilateral.)

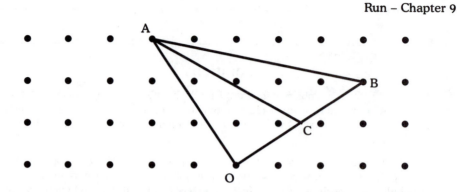

Question 7. Let C be the point of intersection of the altitude from A to the side OB. The segment AC is perpendicular to OB. What can you say about C as a point on the segment OB? Is C a lattice point?

The diagram holds all the information you need to devise a proof, but the algebra can be difficult to manage. Think about how you want to assign coordinates to the points A and C. (You won't need coordinates for B.) Try to argue to the contradiction that $\sqrt{3}$ is a fraction. For some help in choosing coordinates to make the computation more manageable, read the hints below.

Hints: Give A integer coordinates (a, b). C may not be a lattice point, but its coordinates are, at worst, one-half of an integer(half-integers, for short), call them (c, d). Now a and b can be written $a = c + r$ and $b = d + s$, for some integers or half-integers r and s.

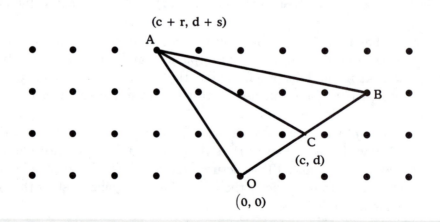

The suggestion is to work with $c + r$ and $d + s$ instead of a and b. This choice will make the computation simpler.

Step 1. Use the chosen coordinates, apply the Pythagorean theorem to the triangle OAC and simplify. The result will be an equation in $c, r, d,$ and s.

Step 2. Try to get 3 into the argument. A suggestion for doing that is to compute the square of the length of the altitude AC in two different ways. One way to do it is to use the coordinates for A and C. Another way is to use the fact that AC is the altitude of an equilateral triangle.

Step 3. Use the equation from Step 1 to substitute for s in the equation in Step 2.

Which proof of the fact that *no equilateral triangle can be constructed in a lattice* do you prefer? Why?

Solutions

Question 7. Let C be the point where the altitude from A intersects the side OB. C is the midpoint of OB since the triangle is equilateral. C may not be a lattice point, but at worst, C has coordinates that are each one-half of an integer (half-integers, for short).

Step 1. If you apply the Pythagorean theorem to the triangle OAC and simplify, you obtain the equation $cr + ds = 0$.

Step 2. You will get $3 = \frac{r^2 + s^2}{c^2 + d^2}$. Since both c and d cannot be 0, you may assume $d \neq 0$.

Step 3. Substitute $s = \frac{-cr}{d}$ into the equation $3 = \frac{r^2 + s^2}{c^2 + d^2}$. You will find that $3 = \frac{r^2}{d^2} = \left(\frac{r}{d}\right)^2$. This is a false statement. Since r and d are either integers or half-integers, the equation says 3 is the square of a fraction, that is, $\sqrt{3}$ is a fraction.

Suggestions for the Endurance Athlete

The Regular Hexagon and Other Regular 3k-gons

10K Challenge. Use the fact that no equilateral triangle is a lattice polygon to verify the following statement.
No regular hexagon can be a lattice hexagon.

20K Challenge. Show that no regular $3k$-gon can be a lattice $3k$-gon.

Regular n-gons

30K Challenge. Investigate the question of which regular polygons can be lattice polygons. Is the square the only regular polygon that can be constructed in a lattice?

References

Coxeter, H. S. M., *Introduction to Geometry*, Wiley, New York, 1961, pp. 208–209.

Jacobs, H. R., *Geometry*, Second Edition, W.H. Freeman & Company, New York, 1987.

Niven, I. and Zuckerman, H. S., "Lattice Points and Polygonal Area," *American Math Monthly* **74** (1967), pp. 1195–2000.

Cut-the-Knot (www.cut-the-knot.com/ctk/Pick.html) is an excellent source for information about Pick's theorem. In particular, you will find a link to a very interesting proof of the theorem. You will also find an interactive geoboard.

Chapter Ten

Dissection

Cutting up, or dissecting, a geometric figure and rearranging the pieces to form another figure is fun. Trying to do it with as few pieces as possible adds to the excitement. This investigation begins with dissection experiments and challenges, and leads to the remarkable discovery that given two polygons having the same area, one can be cut up and its pieces rearranged to form the other.

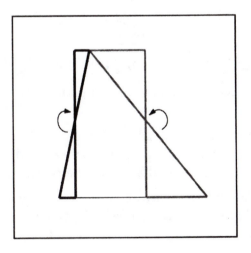

"...[it is] active experience in mathematics itself that alone can answer the question: What is mathematics?"

R. Courant and H. Robbins

Does the prospect of transforming one polygon into another intrigue you? It is not done by wizardry, but by a mathematical procedure called dissection. *Dissection* is the process of taking a plane geometrical object of a particular shape, cutting it up into pieces, and then reassembling all the pieces to make another shape. We will study how polygons such as triangles and parallelograms can be transformed by dissection, or, as we shall say, dissected into other polygons having the same area. We will prove the surprising fact that *every* polygon can be dissected into any other polygon having the same area.

The investigation is designed as a series of challenges. In preparation, here are two warm-up questions.

Question 1. If you take a polygon, cut it into pieces, and reassemble all the pieces to form another polygon, do both polygons have to have the same perimeter? To check this, experiment by cutting a square into pieces.

Question 2. If you take a polygon, cut it into pieces, and reassemble all the pieces to form another polygon, do both polygons have to have the same area?

Part I: Dissecting Rectangles and Right Triangles

Supplies. You will need scissors, a ruler, a protactor or a compass, and some paper.

Preparation. Measure and cut out two $6'' \times 3''$ rectangles, two $9'' \times 2''$ rectangles, and two right triangles each with base $6''$ and height $6''$ where the base and altitude are the sides adjacent to the right angle.

Area Exercise 1. Carefully measure and cut up one of your $9'' \times 2''$ rectangles into as many $1'' \times 1''$ squares as you can. Do you know how many of these $1'' \times 1''$ squares you will get?

Area Exercise 2. Try to reassemble all of the $1'' \times 1''$ squares to form a $6'' \times 3''$ rectangle.

In the area exercise, you showed that a $9'' \times 2''$ rectangle can be dissected into a $6'' \times 3''$ rectangle. Do you think you can do it using fewer pieces? That is your first challenge.

Challenge 1. Find the smallest number of pieces into which you can cut the $9'' \times 2''$ rectangle so that the pieces can be reassembled to form a $6'' \times 3''$ rectangle.

The right triangle with base and height both equal to $6''$ also has area equal to 18 square inches. This suggests the next challenge.

Challenge 2. Find the smallest number of pieces into which you can cut the right triangle so that the pieces can be reassembled to form a $9'' \times 2''$ rectangle.

Solutions

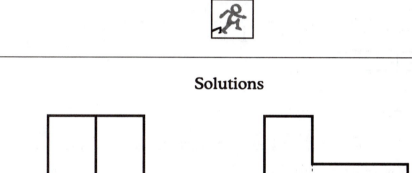

Question 1. The answer is "No." For example, take a $2'' \times 2''$ square and make a vertical cut to form two congruent $1'' \times 2''$ rectangles. Then reassemble them by lying one of the rectangles on its side as in the picture. The square has perimeter $2''+2''+2''+2'' = 8''$. The other polygon has perimeter $2''+1''+1''+2''+1''+3'' = 10''$.

Question 2. The answer is "Yes." This is one of the basic facts about area. If a polygon P is made up of two or more non-overlapping polygons $P_1, P_2, \ldots, P_k$, then the area of P is equal to the sum: area P_1 + area P_2 + ... + area P_k.

Area Exercise 1. Since the area of the $9'' \times 2''$ rectangle is 18 square inches, you should have obtained eighteen $1'' \times 1''$ squares when you cut up the rectangle.

Area Exercise 2. Since the $6'' \times 3''$ rectangle also has area 18 square inches, it is possible to form a $6'' \times 3''$ rectangle with the eighteen $1'' \times 1''$ squares.

Challenge 1. Here is a dissection of the $9'' \times 2''$ rectangle into two pieces to form the $6'' \times 3''$ rectangle.

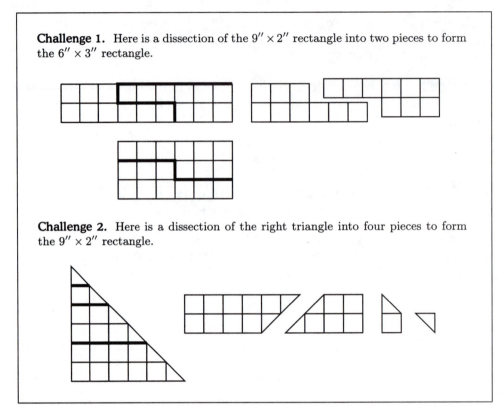

Challenge 2. Here is a dissection of the right triangle into four pieces to form the $9'' \times 2''$ rectangle.

Part II: Dissecting Parallelograms and Triangles

Supplies. You will need scissors, a ruler, a compass or a protractor, tracing paper (or a copier), and paper to make figures.

Preparation. Make, and then cut out, a few copies of the following polygons.

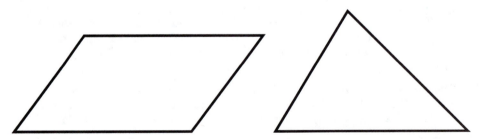

The transformation of a parallelogram, like the one above, into a rectangle is a very simple example of dissection. Can you make this parallelogram into a rectangle with just one cut of the scissors?

Challenge 3. Take one of your copies of the parallelogram. Let b denote the length of its base. Think about how to cut off a piece of the parallelogram so that you can put the two pieces back together to form a rectangle with the same base length b. Then try it.

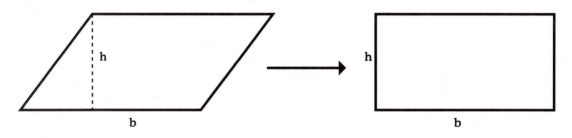

Observe that when you dissect the parallelogram into a rectangle with the same base length, the rectangle will also have height equal to the height of the parallelogram. Why?

Can you undo the procedure? Can you dissect the rectangle with base of length b and height h into a parallelogram with base of length b and height h?

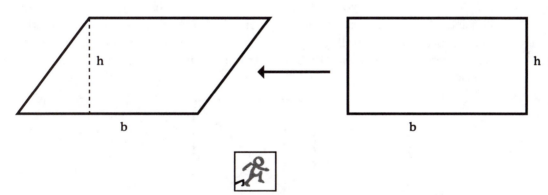

The fact that the dissection of the parallelogram into a rectangle can be reversed to give a dissection of the rectangle into a parallelogram illustrates a very useful idea. Dissection is a *symmetric* procedure. This means that if polygon P can be dissected into polygon Q, then polygon Q can be dissected into polygon P. Just as in the example above, the process can always be reversed.

Challenge 4. Transform the triangle

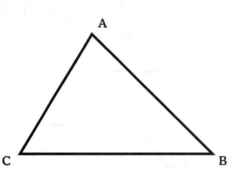

into a rectangle of the same area using dissection.

Start with one of your copies of the triangle. Think about how to cut it into pieces that you can reassemble to form a rectangle. Then try it. Use all the pieces so that the rectangle will have the same area as the triangle.

Hint: Consider how you might use the midpoints of AC and AB.

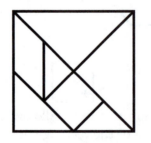

The next challenge concerns The Game of Tangrams, the ancient game based on dissection. Have you ever played tangram games? A set of tangrams consists of seven shapes cut out from one large square. As illustrated below, the shapes are: a small square, two small congruent triangles, two large congruent triangles, a medium sized triangle, and a parallelogram.

There are different ways to play games with tangrams. The idea is to use all the tangram pieces to form various shapes, such as this M (for Mathematics, of course).

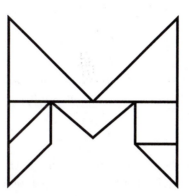

As you will discover, tangram games are dissections of the square into interesting, and, sometimes, unexpected or amusing shapes.

Challenge 5. Cut out your own set of tangrams from a single square sheet of paper.

You can do it by folding and cutting; no ruler is needed. Start out on your own, or follow along if you would like some suggestions.

(If your paper is rectangular make it a square with side length equal to the short side of the rectangle. Do this with one cut. You can measure where to cut by making a diagonal fold of the paper. The leftover rectangle is not needed.)

Step 1. Fold and then cut your square along one diagonal to make two right triangles.

Step 2. Take one of these triangles, and with one fold and cut, make two right triangles. These will be the two "large" triangle tangram pieces.

Step 3. The remaining five tangram pieces will come from the other triangle formed in Step 1. Make the "medium" tangram triangle by folding the right angle corner to meet the hypotenuse at the midpoint. Then cut along this fold. Now you have one "medium" triangle and one large trapezoid.

Step 4. Fold the trapezoid in half so that the slanted edges meet. Cut along this fold.

Step 5. Take one "half-trapezoid" (it's really a whole smaller trapezoid, isn't it?) and, with one fold and cut, make one square and one triangle.

Step 6. Take the other "half-trapezoid," and with one fold and cut, make one parallelogram and a triangle that is congruent to the triangle in Step 5.

This gives you a complete set of tangrams. You know you can form one large square with all the pieces, don't you? Show that the square can be dissected into some interesting shapes of your own invention.

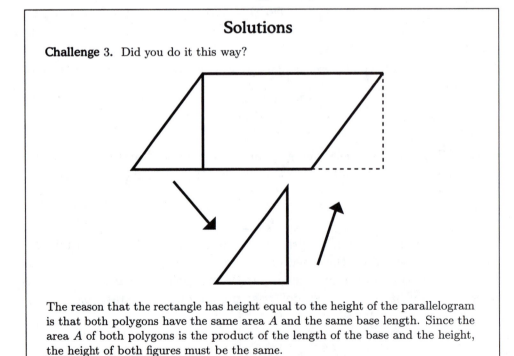

Solutions

Challenge 3. Did you do it this way?

The reason that the rectangle has height equal to the height of the parallelogram is that both polygons have the same area A and the same base length. Since the area A of both polygons is the product of the length of the base and the height, the height of both figures must be the same.

For the dissection of the rectangle into a parallelogram, you have the right idea if you just reversed the process.

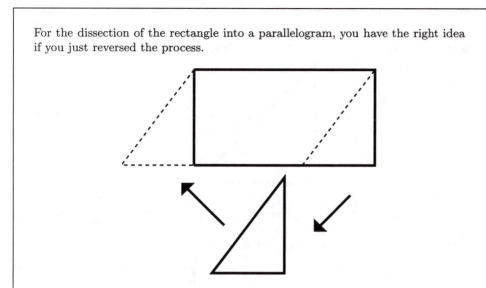

Challenge 4. Let E be the midpoint of AC, and D the midpoint of AB. Drop perpendiculars from E and D to the base BC. Cut the triangle along these line segments. Swing the resulting triangles up to meet at A as in the picture below.

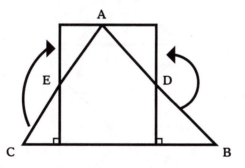

Why does this work? In "Heavy Lifting," you will justify this dissection. You are encouraged to try it now on your own.

Challenge 5. *Step 1.* Fold and then cut your square along one diagonal to make two right triangles.

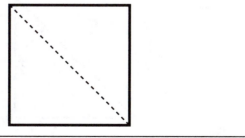

Step 2. Take one of these triangles, and with one fold and cut, make two right triangles. These will be the two "large" triangle tangram pieces.

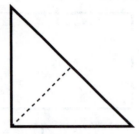

Step 3. The remaining five tangram pieces will come from the other triangle formed in Step 1. Make the "medium" tangram triangle by folding the right angle corner to meet the hypotenuse at the midpoint. Then cut along this fold. Now you have one "medium" triangle and one large trapezoid.

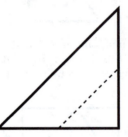

Step 4. Fold the trapezoid in half so that the slanted edges meet.

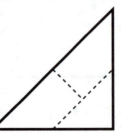

Step 5. Take one "half-trapezoid," and with one fold and cut, make one square and a triangle.

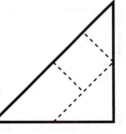

Step 6. Take the other "half-trapezoid," and with one fold and cut, make one parallelogram and a triangle which is congruent to the triangle in Step 5.

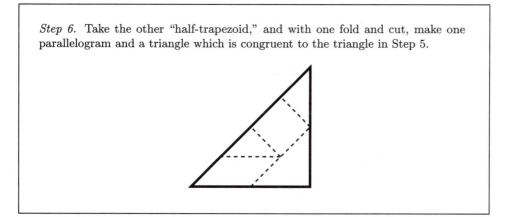

Part III: High Altitude Work-Out

The word "parallelogram" seems automatically to bring the following picture to mind:

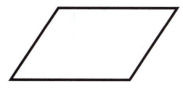

However, there are many types of parallelograms. In this part of our investigation, you will explore the diversity of parallelograms.

Remember that an altitude of a parallelogram is a line segment that joins a vertex on one of the sides to the line of the opposite parallel side (called the base), and is perpendicular to that side. Since each vertex lies on two sides, and each of those sides has an opposite side, there are two altitudes that come from each vertex.

If the foot of the altitude lies on the base inside the parallelogram, we will say the altitude falls *inside* the parallelogram. If the foot of the altitude lies on the line of the base outside the parallelogram, we will say the altitude falls *outside* the parallelogram.

The picture of the parallelogram below shows two altitudes from the same vertex, one falls inside and one falls outside.

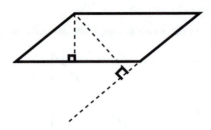

Challenge 6. Draw all the altitudes of the following parallelogram:

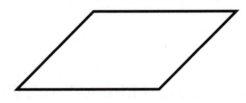

How many altitudes does this parallelogram have? How many fall outside the parallelogram?

It is true that most, but not all, parallelograms that are not rectangles have eight altitudes.

Challenge 7. Draw and count the altitudes of the following parallelogram.

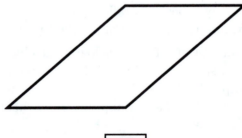

Question 3. How many altitudes does a rectangle have?

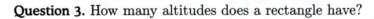

Challenge 8. Next comes a question that challenges our usual notion of a parallelogram. Can a parallelogram have six altitudes that fall outside the figure?

If you are having trouble with this one, try making a skinny and very tilted parallelogram.

Solutions

Challenge 6. Here are the altitudes of the pictured parallelogram:

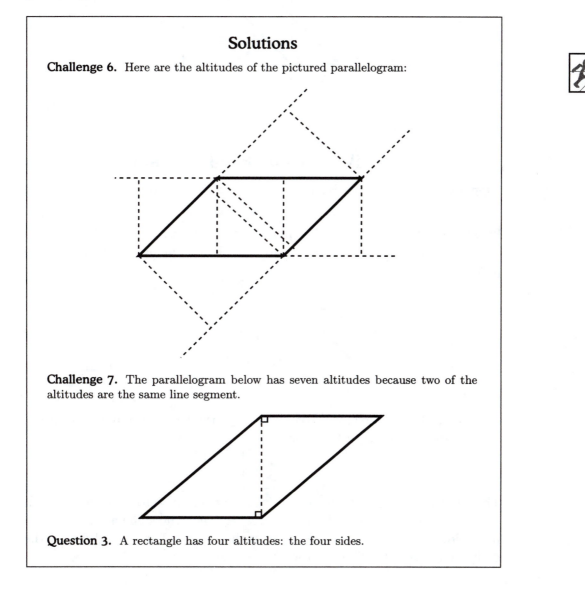

Challenge 7. The parallelogram below has seven altitudes because two of the altitudes are the same line segment.

Question 3. A rectangle has four altitudes: the four sides.

Challenge 8. Here is an example of a parallelogram with six altitudes that fall outside the parallelogram:

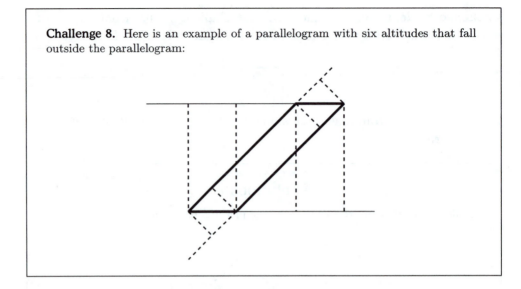

Part IV: Unusual Parallelograms

Supplies. You will need scissors, a ruler, a protractor, and paper to make a parallelogram.

Preparation. Draw a parallelogram with base b of length $2''$ and two parallel sides of length $4''$ that make a $45°$ angle with the line of the base, as in the picture.

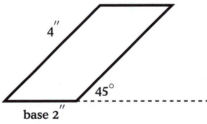

Question 4. If you know some trigonometry, use it to compute the length of an altitude to the line of the base.

The challenge will be to dissect this parallelogram into a rectangle with the lengths of the base and altitude, respectively, the same as those of the parallelogram. First, pause and observe that if one of the $4''$ sides were the designated base, the dissection would be easy.

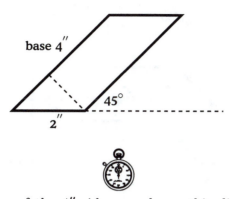

If the base were one of the 4″ sides, as above, this dissection would be a special case of the first dissection in Challenge 3.

The point of the challenge you are going to tackle now is that *the base is the side of length* 2″; you are not allowed to change to the 4″ base. A new strategy will be needed since both altitudes to the 2″ base fall outside the parallelogram.

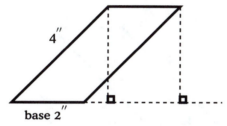

Look at the rectangle that is formed by the top of the parallelogram and the two altitudes to the line of the base.

Question 5. What base length and height does this rectangle have?

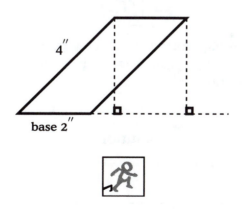

Challenge 9. Cut the parallelogram into pieces and reassemble them to form a rectangle with base length $2''$ and height $2\sqrt{2}''$.

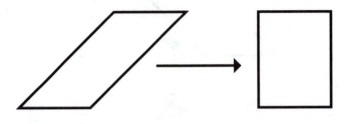

Suggestion 1. Do the dissection in steps. Try to visualize the parallelogram as composed of smaller parallelograms with altitudes falling inside.

Suggestion 2. Put the parallelogram on a piece of paper and draw the altitudes to the line of the base. Make use of the rectangle formed using these altitudes.

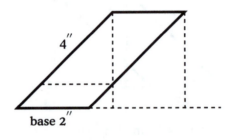

This dissection is an important part of the proof that any polygon can be dissected into any other polygon of the same area. You will do this in the the Heavy Lifting section.

Solutions

Question 4. The parallelogram below has height h equal to $2\sqrt{2}''$, or, approximately, $2.8''$:

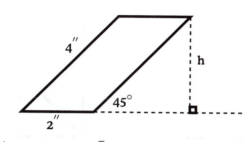

Since $\sin 45^o = \frac{h}{4}$ and $\sin 45^o = \frac{\sqrt{2}}{2}$, it follows that $h = 2\sqrt{2}''$.

Question 5. The rectangle formed by the top of the parallelogram and the two altitudes to the line of the base has base of length $2''$ and height $2\sqrt{2}''$, the same as the original parallelogram.

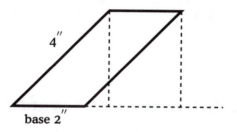

Challenge 9. Here is one way to dissect the parallelogram into a rectangle with base of length $2''$:

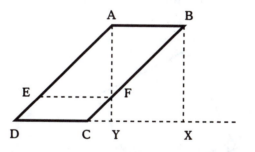

First draw both altitudes to the line of the base. (Place the parallelogram on a piece of paper so that you can draw the altitudes as well as the line of the base.) Both altitudes AY and BX fall outside the parallelogram, but they form with AB and YX the required $2'' \times 2\sqrt{2}''$ rectangle.

Next, make one cut, parallel to the base, along EF. Now you have two parallelograms.

Cut $ABFE$, the larger parallelogram, along the left-hand altitude, AF, to form two triangles. Slide the left-hand triangle AFE so that it meets the right-hand triangle ABF along its hypotenuse to form triangle BGF. Look what you have—most of the required rectangle.

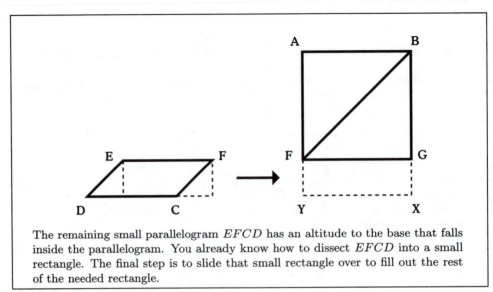

The remaining small parallelogram $EFCD$ has an altitude to the base that falls inside the parallelogram. You already know how to dissect $EFCD$ into a small rectangle. The final step is to slide that small rectangle over to fill out the rest of the needed rectangle.

Part V: The Pythagorean Theorem by Dissection

In this part of the investigation, you will use dissection to prove the most famous (and most proved) mathematical theorem of all, the Pythagorean theorem. The picture below illustrates the theorem.

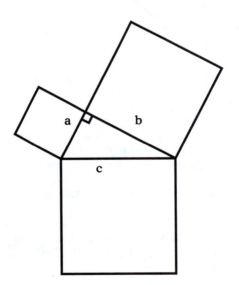

Pythagorean Theorem. *If squares are constructed on the three sides of a right triangle, the area of the square on the hypotenuse is equal to the sum of the areas of the squares on the legs. If the legs of the right triangle have lengths a and b, and if c is the length of the hypotenuse, then*

$$c^2 = a^2 + b^2.$$

Preparation. The dissection proofs that you are going to investigate show that the Pythagorean theorem is true for any right triangle. But it will be fun also to do hands-on demonstrations of the proofs for a particular right triangle. So, to prepare, carefully construct some figures. (For assistance in constructing right angles and reproducing angles and lengths using a compass, see the Solutions.)

First, construct a right triangle: Draw a horizontal line segment with left-hand endpoint C. Then, using whatever tools you choose, such as a compass or protractor, draw a line segment making a right angle at C.

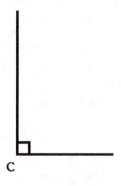

Join the endpoints of the two segments to form a right triangle. Call the lengths of the legs of your triangle a and b, and the length of the hypotenuse c.

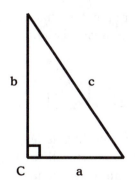

Cut out four copies of the triangle.

Next, using your tools (you can use, for instance, a compass to reproduce lengths), construct and cut out four squares, one each having side length equal to a, b, c and, finally, $a + b$.

Just for fun, reproduce the balancing act in the Pythagorean picture above with some of your cut-out figures.

Each of the two challenges below provides a proof by dissection of the Pythagorean theorem.

Caution. Drawing pictures of dissections, or actually cutting up figures and reassembling the pieces does not constitute a complete proof. The pictures must be backed up with facts. Often, it is expected that the readers will furnish the necessary facts. Try to do this. (For some very convincing examples of fake dissections, see the web references at the end of this section.)

Challenge 10. Part 1. Dissect your big square, the one with side length equal to $a + b$, into 4 copies of your triangle and one square with side length equal to c, the length of the hypotenuse of your right triangle.
Part 2. Explain why the dissection in Part 1 proves the Pythagorean theorem.

To carry out Part 1 of this challenge, start with the big square. Then figure out how to arrange four copies of your right triangle and your square with side length c on top of the big square so they fit exactly with no overlaps.

For Part 2 of the challenge, remember that the area of the big square must be equal to the sum of the areas of all the polygons in any dissection of it, and use a little algebra.

Note. The picture of the dissection in Challenge 10, which is found in the Solutions, should be familiar to you. Recall the construction of lattice squares where the side length s is a sum of two squares of positive integers. Sums of squares play an important role in mathematics. The Pythagorean theorem is one of the earliest theorems about sums of two squares.

The next challenge will give a different dissection of the big square. Then, using both dissections, you will obtain another proof of the Pythagorean theorem without using any equations at all.

Challenge 11. Find a second dissection of your big square, the one with side length equal to $a + b$. This time use four copies of your triangle, your square of side length equal to a, and your square with side length equal to b. Then compare this dissection with the one from Challenge 10 to produce a proof of the Pythagorean theorem.

To carry out this challenge, start with the big square. Then figure out how to arrange four copies of your right triangle, your square with side length a, and your square with side length b on top of the big square so they fit exactly with no overlaps.

Another, quite different, proof of the Pythagorean theorem by dissection can be found in the Heavy Lifting section.

Application of the Pythagorean Theorem

Question 6. Think about the following question. Do you know how to construct a square with side length that is not an integer and not a fraction, say a square with side length exactly equal to $\sqrt{2}$? (No approximations, please.)

If you are not sure how to do this, consider whether the Pythagorean theorem will provide some help. All you need to start building a square of side length equal to $\sqrt{2}$ is one line segment of length equal to $\sqrt{2}$. Can you find a method for constructing a line segment of length $\sqrt{2}$?

Question 7. Why stop with $\sqrt{2}$? Now that you know how to construct a segment of length $\sqrt{2}$, can you construct one of length $\sqrt{3}$? How about

segments of length $\sqrt{5}, \sqrt{6}, \sqrt{7}$, and $\sqrt{8}$? Can you construct them? If you can, then you can then use a compass to reproduce lengths and construct squares having these irrational numbers as side lengths.

Excellent work. If you are eager to tackle some dissection problems with a higher degree of difficulty, continue on to "Heavy Lifting."

Solutions

Preparation. For some assistance in constructing right angles and reproducing angles and lengths using a compass read this.

To duplicate a given line segment AB:

Step 1. Draw a line l and mark a point A' on it.

Step 2. Set the radius of the compass equal to the length of AB by putting the metal point at A and the pencil point at B.

Step 3. With the radius set as above and with the metal point of the compass at A', draw an arc that intersects l. Call the point of intersection B'.

Then segments AB and $A'B'$ have the same length.

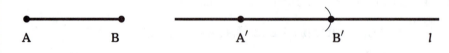

To duplicate a given $\angle A$:

Step 1. Draw a ray r emanating from a point A' as one side of the angle to be constructed.

Step 2. Using the compass, draw an arc that intersects the sides of $\angle A$ at points B and C.

Step 3. Keeping the radius of the compass fixed and with the metal point at A', draw an arc that intersects the ray r at a point C'.

Step 4. Set the radius of the compass equal to the distance between B and C.

With the metal point at C', draw an arc that intersects the previous arc and let B' be the point of intersection of the two arcs. Draw segment $A'B'$.

Then the measure of angle $\angle CAB$ is equal to the measure of angle $\angle C'A'B'$.

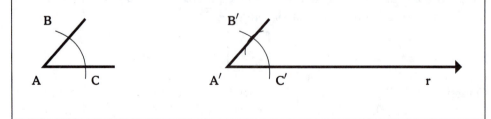

To construct the perpendicular bisector of a segment AB:

Step 1. With A and then B as centers, draw two arcs that have the same radius and intersect each other at points C and D. Draw CD.

Step 2. CD is the perpendicular bisector of the segment AB.

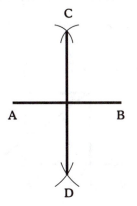

Challenge 10. Part 1. Here is a dissection that answers the challenge.

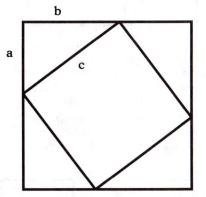

It is necessary to give some justification of the fact that the figure formed above by the hypotenuses of the four copies of the right triangle (with legs of length a and b and hypotenuse of length c) is a square. It is clear that the figure is a quadrilateral with four sides of equal length. It follows from the congruence of the right triangles and from the fact that the sum of the measures of the angles in a triangle is equal to $180°$ that the sides of the quadrilateral triangle meet at right angles.

Part 2. The area of the big square is $(a+b)^2 = a^2 + 2ab + b^2$. It must be equal to the sum of the areas of all the pieces in the dissection. So, using the fact that the area of each right triangle is $\frac{1}{2}ab$ and that there are four of them, the following equation is obtained:

$$c^2 + 4(\frac{1}{2}ab) = a^2 + 2ab + b^2,$$

$$c^2 + 2ab = a^2 + 2ab + b^2.$$

Now, if you subtract $2ab$ from both sides of this equation, the Pythagorean equation

$$c^2 = a^2 + b^2,$$

results.

Challenge 11. Here is a dissection that answers the challenge.

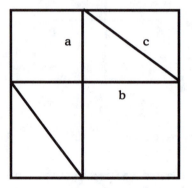

For this dissection, it is straightforward to justify that the right triangles and squares fit together as claimed by the picture.

We use both dissections to prove the Pythagorean theorem. Since you found two dissections of the same square, the sum of the areas of all the pieces in the first dissection must be equal to the sum of the areas of all the pieces in the second dissection. So, since the same four triangles are part of each dissection, they cancel each other. Therefore, the area of the square of side c must be equal to the sum of the areas of the squares of sides a and b.

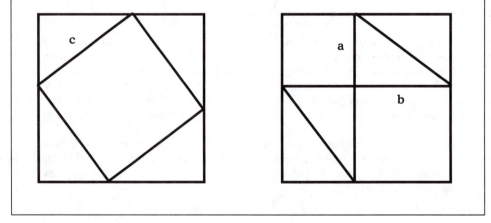

Question 6. If you start by constructing a right angle having both legs of length 1, then the line segment that completes the triangle has, by the Pythagorean theorem, length $\sqrt{2}$. Finally, you can use a compass to reproduce this length and construct a square of side length $\sqrt{2}$.

Question 7. In the application of the Pythagorean theorem, you are asked how you construct segments of lengths $\sqrt{3}, \sqrt{5}, \sqrt{6}, \sqrt{7}$, and $\sqrt{8}$. There are many ways to do this. Here is one. It uses the fact that you know how to construct a segment of length $\sqrt{2}$. Write $3 = 1 + (\sqrt{2})^2$. You know how to construct a segment of length $\sqrt{2}$, so construct a right triangle with legs of lengths 1 and $\sqrt{2}$. By the Pythagorean theorem, its hypotenuse will have length $\sqrt{3}$. Note that $3 = 1 + (\sqrt{2})^2$ expresses 3 as the sum of squares of two numbers that are not necessarily integers. This differs from the sum of two squares discussion in Chapter 9.

Now you know how to construct line segments of length $\sqrt{2}$ and $\sqrt{3}$. You can construct the remaining segments as follows. First write the numbers $5, 6, 7$, and 8 as a sum of two squares: $5 = 1 + 2^2, 6 = 2^2 + (\sqrt{2})^2, 7 = 2^2 + (\sqrt{3})^2$, and $8 = 2^2 + 2^2$. Next, use the method above to construct segments of length $\sqrt{5}, \sqrt{6}, \sqrt{7}$, and $\sqrt{8}$.

References

Jacobs, H. R., *Geometry*, Second Edition, W. H. Freeman & Company, New York, 1987.

If you use a search engine such as Google, the word "tangram" will call up many references such as http://mathforum.org.trscavo/tangrams.html.

For animated tangrams, see www.tygh.demon.co.ek/tan.

Take a look at www.cut-the-knot.com/pythagoras/tricky.html for an interesting fake dissection.

The site www1.ics.uci.edu/∼ eppstein/junkyard/dissect.html has references to more fake dissections.

Part VI: Heavy Lifting

Are you ready to put your dissection talents to work to transform any polygon into any other polygon of the same area? The proof of this amazing fact (known in the literature as the Bolyai-Gerwien theorem) will require some hard work and ingenuity on your part. We will end with some enjoyable applications of dissection, including a third, very clever, dissection proof of the Pythagorean theorem.

At this point in the investigation, the earlier dissection of a $9'' \times 2''$ rectangle into a $6'' \times 3''$ rectangle seems very easy. The same cannot be said about the dissection of a $9'' \times 2''$ rectangle into a $3\sqrt{2}'' \times 3\sqrt{2}''$ square. That dissection is quite a bit more challenging. It will be your first project in this section. First, be sure you know how to construct a $3\sqrt{2}'' \times 3\sqrt{2}''$ square. Do that now.

Did you first construct a right triangle with both legs of length $3''$ and use the Pythagorean theorem?

The proof that any polygon can be dissected into any other having the same area will be the result of the two projects in this section. The first project is the proof that every rectangle can be dissected into every other rectangle having the same area. You are encouraged to make demonstration models as you go along, for the proof is very interesting and intricate.

Project 1

> Every rectangle can be dissected into any other
> rectangle having the same area.

To get from one rectangle to the other one we have to go by way of a parallelogram. The first step is to apply the technique for any parallelogram that you used earlier in Part IV for a special parallelogram.

Step 1. Every parallelogram can be dissected into a rectangle having the same base length b and altitude length h.

You might think "I did this already!" It is true that you have already executed the dissection in some special cases, but now you have to show that the statement is true for an arbitrary parallelogram. In Part II, you dissected a parallelogram to transform it into a rectangle, but it was a parallelogram in the "standard form": one of the altitudes to the base fell inside the parallelogram. In Part IV, you did the dissection for a specific parallelogram having both altitudes to the base falling outside the parallelogram. The proofs in these special cases will be helpful in the development of the proof of the general case.

It is always useful to have a demonstration model. Draw a parallelogram with base of length $\sqrt{3}''$ and sides of length $4''$ making an angle of 30° with the line of the base.

Question 8. What is the length of an altitude to the base?

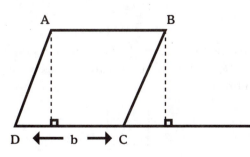

Demonstration model:

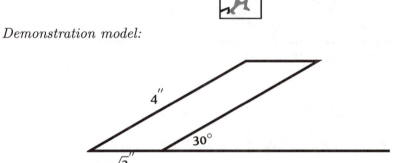

As you figure out a proof for Step 1, try it out on your demonstration model. This model will illustrate more of the complexities of the proof than the one used earlier in Part IV.

If you want to prove Step 1 on your own, think about the strategy used in Part IV and try to make it work for any parallelogram.

Otherwise, use the following outline for the proof. (Even with the outline, there will still be plenty of work for you to do!)

Let the parallelogram have vertices $A, B, C,$ and D. Draw both altitudes to the line of the base CD. There are two cases. You can call them the *in-case* and the *out-case*, depending on whether one of the altitudes to the base falls inside the parallelogram or not.

(a). The diagram below shows an example of the *in-case*, since the foot of the altitude from A to the base CD lies inside the parallelogram.

(b). An example of the *out-case* is pictured below. The foot of each altitude to the base CD lies outside the parallelogram.

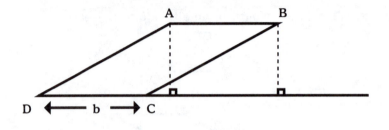

The demonstration model is an example of an out-case.

Challenge 12(a). For the proof of Step 1, do the easy in-case first.

In-case: If one of the altitudes to the base CD falls inside the parallelogram

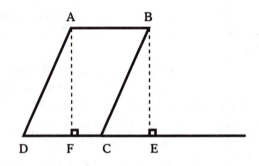

you know how to transform it by dissection into a rectangle with the same base length b. Do it now.

Did you decide to slide triangle AFD on top of triangle BEC? To justify this step, you must show that the two triangles are congruent. Did you do that? If not, try to do it now.

Observe that an extreme case occurs when the foot of an altitude to the base lies at a vertex of the base as in the following picture:

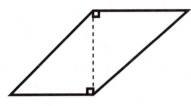

We will call this case the *extreme in-case*.

Challenge 12(b). Tackle the out-case where both altitudes fall outside the parallelogram. The idea here is to reduce the out-case to the in-case.

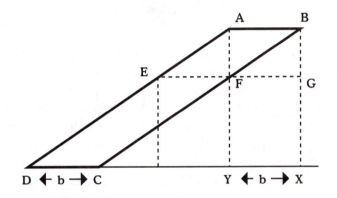

Out-case: The two altitudes AY and BX together with AB and a segment XY of equal length on the line of the base CD form the rectangle $ABXY$. Rectangle $ABXY$ has the same base length b and height h as parallelogram $ABCD$. Consider trying the method you used in the dissection in Part IV. There you visualized the parallelogram as composed of several in-case parallelograms with the same base length b.

If you apply the method suggested in the solutions and dissect the extreme in-case parallelogram $ABFE$ into rectangle $ABGF$, you are left with $EFCD$, a parallelogram that is part of the original parallelogram and has the same base.

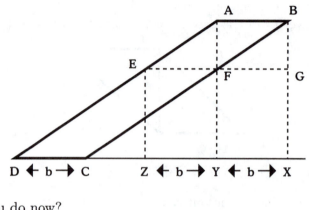

What do you do now?

You repeat the procedure on parallelogram $EFCD$. One altitude may fall inside parallelogram $EFCD$, so you have the easy in-case, and one more step finishes the proof. (This is what happened with the parallelogram in Part IV.) Otherwise, you have the out-case with both altitudes falling outside the parallelogram. You must repeat the procedure that slices at the point of intersection of the left hand altitude and the right hand side to produce an extreme in-case parallelogram. This is what happens in the demonstration model, isn't it? The idea is the skinnier and more tilted the parallelogram, the more repetitions needed. The picture looks like this:

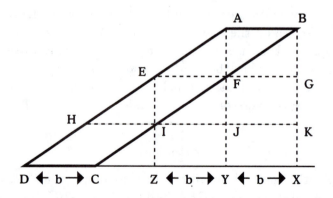

Now slide triangle EIH on top of triangle FJI, as before, then slide all of rectangle $EFJI$ on top of rectangle $FGKJ$. You have covered a larger part of the big rectangle, and you are left with an even smaller parallelogram $HICD$ on the same base CD. What do you do now?

You are right if you said "Repeat the procedure!" In the demonstration model, at this stage one altitude falls inside the parallelogram so it is the easy in-case, and one more step finishes the proof. For the general parallelogram, you might need to repeat more times.

Question 9. Can you see why eventually you will get a parallelogram with the left-hand altitude falling inside the parallelogram?

This completes the proof that every parallelogram can be dissected into a rectangle having the same base length b and height h. Now check the Solutions for Step 1, and then try Step 2.

Solutions

Question 8. The parallelogram with base of length $\sqrt{3}''$ and sides of length $4''$ making an angle of $30°$ with the line of the base has height $2''$. This follows from the fact that $\sin 30° = \frac{1}{2}$.

Challenge 12(a). *In-case.* To dissect parallelogram $ABCD$ into rectangle $ABEF$,

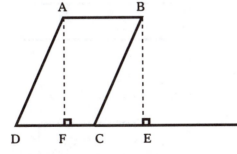

slide triangle AFD on top of triangle BEC. To justify this step, the two triangles must be proved congruent. There are many ways to do this. For example, you know the parallelogram's opposite sides AD and BC have the same length and its altitudes AF and BE have the same length, and so, since the triangles are right triangles, the Pythagorean theorem applies and the triangles are congruent by side-side-side.

Challenge 12(b). *Out-case:*

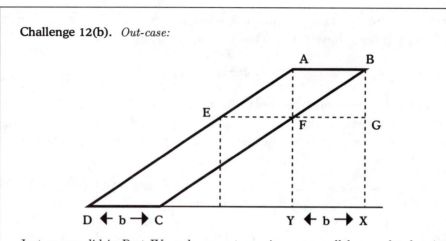

Just as you did in Part IV, make an extreme in-case parallelogram by drawing the line EF parallel to the base, where F is the point of intersection of the left altitude AY with the right side BC of the parallelogram. Since parallelogram $ABFE$ with altitude AF is the extreme in-case, and you just proved that triangle AFE is congruent to triangle BGF, you can slide triangle AFE on top of triangle BGF to get part of the big rectangle.

Question 9. Each time you have to repeat the process, the points $Y, Z, \ldots$ at the foot of the left hand altitude are all the same distance b apart, and they move closer and closer to D. This means that eventually one will lie between C and D, perhaps at C itself.

Step 2. Every rectangle can be dissected into any other rectangle having the same area.

Demonstration models:

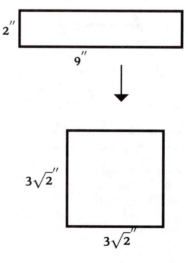

To help follow the steps of the proof, cut out two demonstration models, a $9'' \times 2''$ rectangle, and a $3\sqrt{2}'' \times 3\sqrt{2}''$ square. Then, using the proof below as a guide, cut up the $9'' \times 2''$ rectangle, and reassemble the pieces into the $3\sqrt{2}'' \times 3\sqrt{2}''$ square.

Step 2 starts with two rectangles both having the same area R. We have some freedom of choice in notation. Let one rectangle have base of length x and altitude of length y with $x \geq y$, and the other have base of length w and altitude of length z with $z \geq w$.

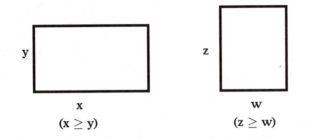

$$y \quad\quad\quad x \quad\quad\quad z \quad\quad w$$

Since both rectangles have the same area, we have $R = xy = wz$. If both rectangles have the same dimensions, no dissection is required. This happens when $x = z$ or $x = w$, in which case $y = w$ or $y = z$, respectively. So the assumption that $x \neq z$, and therefore, that $y \neq w$ can be made. In particular, $y < w$ or $y > w$.

Question 10. Can you put $x, y, z,$ and w in order in the case where $y < w$?

If $y > w$, the roles of x and y can be interchanged. We may assume, therefore, that the relation between x, y, z and w for Step 2 is:

$$x > z \geq w > y.$$

The idea of the proof is to go from one rectangle to the other via a parallelogram. First, the rectangle with base of length x and height y is transformed by dissection into a parallelogram with base of length x and height y. Then Step 1 is used to transform this parallelogram into a rectangle with base of length w and height z.

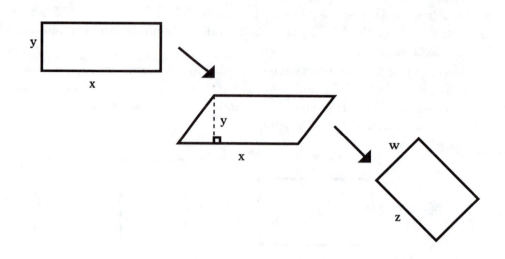

You are going to work out the proof for any x, y, z, and w, but it will be helpful to use the demonstration models as guides. For the demonstration models, let the $9'' \times 2''$ rectangle be the rectangle with base of length x and height y. So $x = 9''$ and $y = 2''$. Let the $3\sqrt{2}'' \times 3\sqrt{2}''$ square be the rectangle with base of length w and height z. So $w = z = 3\sqrt{2}''$.

First you will transform the $9'' \times 2''$ rectangle by dissection into a parallelogram with base of length $9''$ and altitude of length $2''$. Then you will transform this parallelogram into the $3\sqrt{2}'' \times 3\sqrt{2}''$ square.

This picture will help with the proof. The rectangle $ABCD$ has base CD of length x and altitude AD of length y.

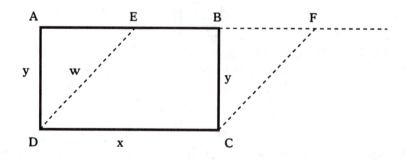

Since $x > w > y$, there is a point E between A and B so that DE has length w. Now construct CF parallel to DE, so that $EFCD$ is a parallelogram. Since $EFCD$ has base CD and altitude BC, it follows that $EFCD$ has area $R = xy = wz$.

Challenge 13. Begin by dissecting rectangle $ABCD$ into parallelogram $EFCD$.

The next step contains a very clever idea: Look at parallelogram $EFCD$ in another way. Think of DE as the base. Draw altitudes DH and EK from D and from E, respectively, to the line of CF.

Here are pictures of two possible arrangements that might arise.

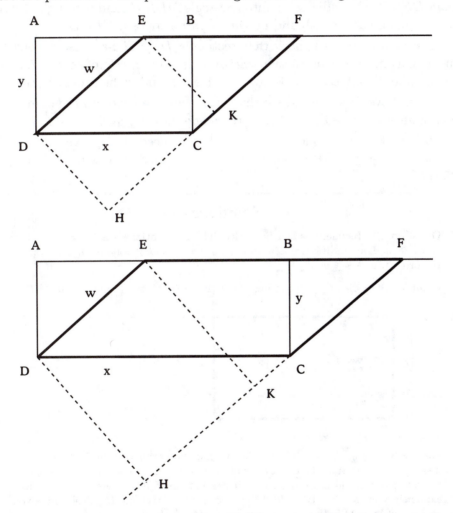

In the top figure, the altitude from E to the line of CF falls inside the parallelogram $EFCD$; in the bottom figure, the altitudes from E and D to the line of CF fall outside the parallelogram.

Question 11. You know parallelogram $EFCD$ has area $R = xy = wz$. When you compute the area of parallelogram $EFCD$, thinking of it as having base DE of length w, what can you conclude about the length of the altitudes DH and EK?

Now, here is the payoff for your hard work in Step 1. By Step 1, parallelogram $EFCD$ can be dissected into rectangle $DEKH$, and rectangle $DEKH$ has the same base length and height as parallelogram $EFCD$.

The final step is to check that rectangle $DEKH$ has base of length w and height z. By construction, $DEKH$ has base DE of length w. You argued above that DH has length z. This concludes the proof of Step 2.

If you haven't already done this, carry out the proof on the demonstration models to dissect the $9'' \times 2''$ rectangle into the $3\sqrt{2}'' \times 3\sqrt{2}''$ square.

Project 1 contains most of the hard work needed to prove the amazing result in Project 2. Proceed to Project 2 and reap the rewards of your efforts.

Solutions

Question 10. Inequalities can be tricky. The assumptions are: $xy = zw$, $x \geq y$ and $z \geq w$. If in addition, $y < w$, then, since $xy = zw$, it follows that $x > z$. So the order is $x > z \geq w > y$.

Challenge 13. Begin by dissecting rectangle $ABCD$ into parallelogram $EFCD$.

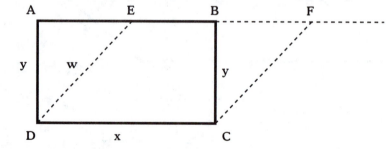

Since $x > w > y$, the altitude BC lies within the parallelogram $EFCD$, so the dissection is easy. It is just the reverse of the in-case dissection from parallelogram to rectangle you did in Step 1. That it can be done follows by symmetry. More explicitly, you can slide triangle AED over on top of triangle BFC, since (as you checked in Step 1) both triangles are congruent.

Question 11. Since parallelogram $EFCD$ has area $R = xy = wz$, and has base DE of length w, it follows that the length of the altitudes DH and EK is z.

Project 2

> If P and Q are polygons having the same area,
> then P can be dissected into Q.

Just follow the steps. The proof will use facts from Project 1.

Step 1. Triangulation. Every polygon can be dissected into triangles all of whose vertices are vertices of the original polygon.

Challenge 14. Can you prove this statement if the polygon is convex? Remember that a polygon is convex if every line segment joining any pair of vertices lies within the polygon.

Suppose that the convex polygon has n sides. If $n = 3$, the polygon is a triangle, so there is nothing to prove. Now assume that $n > 3$. How can you cut the polygon up into triangles, whose vertices are those of the original polygon?

Challenge 15. It is also true that non-convex polygons can be triangulated. Can you triangulate this polygon?

It is true for any polygon, convex or non-convex, that there always is a pair of vertices that can be connected by a line segment that lies wholly inside the polygon (except for its endpoints). This is easy to prove for convex polygons, but more difficult for non-convex polygons. You might make the non-convex case a future project.

Step 2. Every triangle can be dissected into a rectangle.

You did this in Part II. But now, in the more rugged part of the exercise, you must use geometry to justify the dissection you executed there.

Rotate the triangle, if necessary, so that the largest angle is up at the top, at *A*. Then the altitude to the base *BC* will lie inside the triangle.

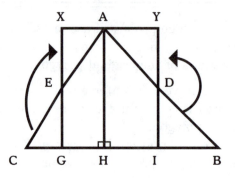

Draw *AH*, the altitude from *A* to the base *BC*. Let *D* be the midpoint of *AB*, and *E* the midpoint of *AC*. Construct lines through *D* and *E* that are parallel to *AH*. Let *X* and *Y* be the points of intersection of these lines with the line through *A* parallel to *BC*. If you show that triangles *EGC* and *EXA* are congruent, and that triangles *DBI* and *DAY* are congruent, then you can construct rectangle *XYIG* by rotating triangles *EGC* and *DBI* as indicated.

Challenge 16. Show that triangles *EGC* and *EXA* are congruent, and that triangles *DBI* and *DAY* are congruent.

Step 3. Every rectangle can be dissected into any other rectangle of the same area.

You proved this in Project 1.

Step 4. If polygons *P* and *Q* of area *R* can be dissected into a square of area *R*, then *P* can be dissected into *Q*.

Can you figure out why this is true?

Here is one way to see it. Cut up polygon *Q* into pieces and reassemble them to make the square *S* of area *R*. Now imagine that polygon *P* is drawn on tracing paper. Cut up polygon P, and reassemble the pieces to fit exactly on top of the square S made from the pieces of Q. Next, look through the tracing paper to the pieces of Q underneath. Use them as a guide to cut some of the pieces of P into smaller pieces, if necessary, and then put all of these pieces together to form Q. This process allows us to dissect a polygon P into a square S, refine the dissection of S, and reassemble it into a polygon Q, illustrating the transitive property of dissection.

For example, carry out the above procedure on the following polygons P and Q.

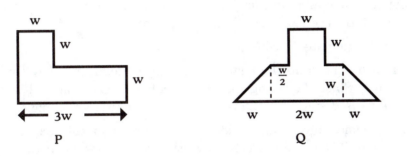

First, dissect Q into a square of area R.

Next, dissect P and reassemble the pieces on top of the square formed from Q.

Finally, look through the top square to observe how you have to cut up some pieces of P, and put them together to form Q.

> All that remains is to put all steps together to show that any polygon P of area R can be transformed by dissection into a square of area R.

Putting All the Steps Together for the Proof

Suppose you prove that any polygon P of area R can be dissected into a square of area R. Then the same will be true for any other polygon Q of area R. So by Step 4, polygon P can be dissected into polygon Q.

We will prove that any polygon P of area R can be dissected into a square of area R. This is what you are going to do. To begin, you will dissect P into a finite number of triangles. Next, you will dissect each triangle into a rectangle of the same area. Then, you will dissect that rectangle into one having the same area and having one side of length $s = \sqrt{R}$. Finally, you will reassemble the rectangles into a square of area $R = s^2$.

1. First, triangulate P using Step 1.

2. For each triangle in the triangulation do the following: Say the triangle has area A. Since the triangle is part of polygon P, the area A is less than or equal to R. Using Step 2, dissect the triangle into a rectangle of area A.

3. Here comes a very clever idea. The area A of this rectangle is not bigger than $R = s^2$. So A can be written as s times "something," where "something" is at most equal to s. This means that there is a rectangle of area A that has a side of length s. The stage is set for you to use Project 1 to dissect the rectangle into another rectangle, one of whose sides has length s.

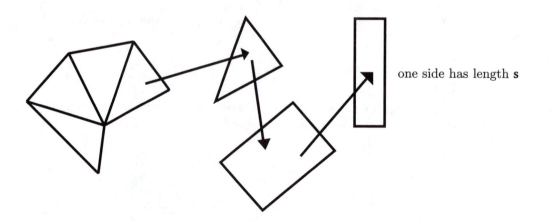

one side has length **s**

4. After you have transformed all of the triangles in the triangulation into rectangles with one side having length s, assemble all of them together placing sides of length s next to one another.

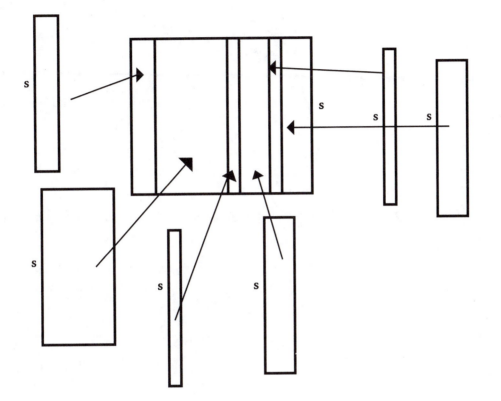

The resulting big rectangle is composed of *all* the pieces involved in *all* the dissections in the earlier steps, so it has area $R = s^2$. Since one of the sides of this rectangle has length $s = \sqrt{R}$, the other side must also have length s. Thus, the big rectangle is a square. Your construction of a square of area R is complete.

Don't you agree that this is a very interesting proof of a very interesting theorem?

Solutions

Challenge 14. There are many ways to triangulate a convex polygon. Did you do something like this? If the vertices are labeled $P_1, P_2, \ldots, P_n$, cut along the line segments $P_1 P_3, P_1 P_4, \ldots, P_1 P_{n-1}$, which all lie inside the polygon.

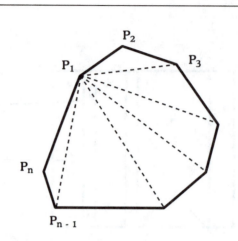

Challenge 15. The given non-convex polygon can be triangulated as follows.

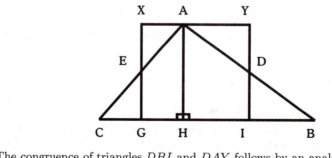

Challenge 16. Triangles EGC and EXA are right triangles. The lengths of AE and EC are the same. AC is transversal for parallel lines XY and BC, so the measures of $\angle XAE$ and $\angle GCE$ are the same. Thus the measures of $\angle AEX$ and $\angle CEG$ are the same. This proves that the triangles EGC and EXA are congruent by angle-side-angle.

The congruence of triangles DBI and DAY follows by an analogous argument.

Step 4. There are many ways to carry out this illustration of Step 4. Here is one. Polygon Q can be dissected into a square as follows:

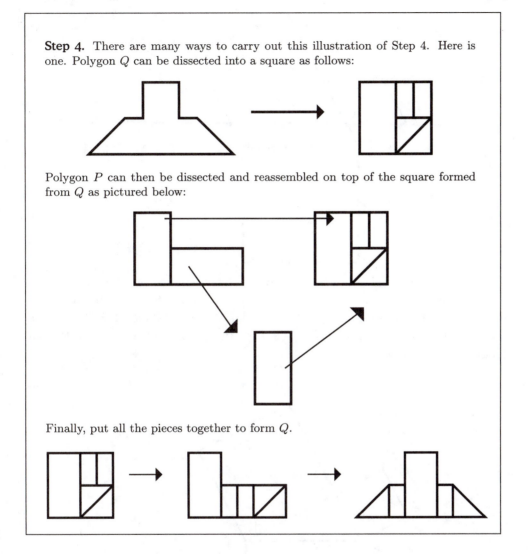

Polygon P can then be dissected and reassembled on top of the square formed from Q as pictured below:

Finally, put all the pieces together to form Q.

Interesting Applications of Dissection

Here are some enjoyable applications of dissection that you can regard as cool-down exercises after the strenuous work on the polygon dissection theorem.

Application 1. (Yet another) proof of the Pythagorean theorem. To begin, draw the Pythagorean picture.

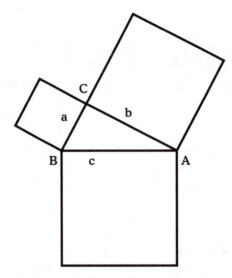

The idea behind this proof is to show that the square on the hypotenuse dissects into two rectangles, one having area a^2 and the other having area b^2. Be sure to justify each step. First, let the extensions of FG and HI meet at W. Quadrilateral $GWHC$ is a rectangle and its diagonal CW has length c.

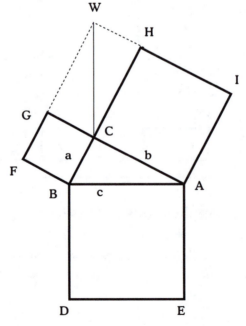

Construct parallelograms $WCBT$ and $WYAC$. Note that BT extends BD and AY extends AE.

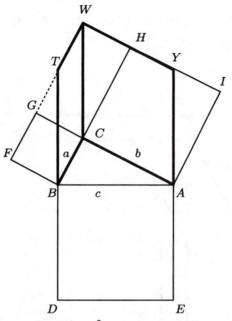

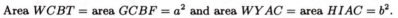

Area $WCBT$ = area $GCBF = a^2$ and area $WYAC$ = area $HIAC = b^2$.

Slide parallelograms $WCBT$ and $WYAC$ vertically downward so that WT meets BC and WY meets AC. Let C' be the image of C.

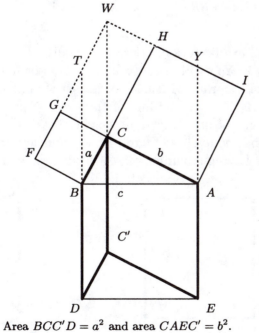

Area $BCC'D = a^2$ and area $CAEC' = b^2$.

Let R be the point of intersection of WC' with AB and let S be the point of intersection of the extension of WC' with DE.

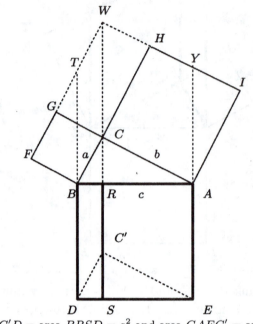

Area $BCC'D =$ area $BRSD = a^2$ and area $CAEC' =$ area $RAES = b^2$.

Thus $a^2 + b^2 = c^2$.

Application 2. Roll-up dissection. Here is a really fun way to do the dissection in Project 1. You will need a ruler, protractor, paper, scissors, and tape. Draw a parallelogram with base of length $4''$ and sides of length $8''$, making an angle of $45°$ with the line of the base. Next, draw a rectangle with the same length of base and same height. It's a $4'' \times 4\sqrt{2}''$ rectangle, right? Cut out the rectangle, and roll it up from left to right to form a cylinder with base having circumference $4''$. Put tape along the seam.

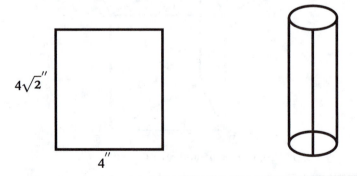

Now, cut out the parallelogram, and place the bottom left vertex of the parallelogram on top of the cylinder, as in the picture below:

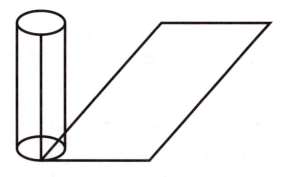

Next, wind the parallelogram onto the tube by matching the base of the parallelogram with the circumference of the base of the cylinder.

Since both polygons have the same height, the parallelogram will climb up the cylinder with no overlaps, and the side opposite its base will correspond exactly with the top edge of the cylinder.

Very carefully, tape the sides of the parallelogram together. You now have two cylinders, one made by the parallelogram on top of the one made made by the rectangle. Slip off the top cylinder, and cut vertically downward from the top right vertex to the base of the cylinder. Flatten it, and look at the *outside* surface of the resulting rectangle. What do you see?

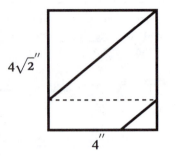

On the surface of the rectangle, you see the steps in the dissection done earlier in Part IV. (To facilitate the construction, the model used here is larger but similar to the one used in Part IV.) You can see that two steps were needed. The first step is an out-case to extreme in-case reduction. The second step is an easy in-case. Neat, don't you think?

Try this proof on the demonstration model parallelogram with base of length $\sqrt{3}''$ and height $2''$ from Step 1 of Project 1. That dissection required two reductions of out-case to extreme in-case, and one in-case. Since the in-case step produces a very small triangle, you will have to be very precise or else you will miss it!

Suggestions for the Endurance Athlete

10K Challenge. Prove that *every* polygon, convex or not, can be triangulated. You might begin by proving that, for any polygon, convex or non-convex, there always is a pair of vertices that can be connected by a line segment lying wholly (except for its endpoints) inside the polygon.

References

Levy, L. S., *Geometry: Modern Mathematics via the Euclidean Plane*, Prindle, Weber and Schmidt, 1970, Chapter 5, "Polygonal Dissection."

Frederickson, G. N., *Dissections: Plain and Fancy*, Cambridge University Press, 1997.

The site www.mathworld.wolfram.com/Dissection.html displays minimal dissections of one regular polygon into another.

The site www1.ics.uci.edu/~eppstein/junkyard/dissect.html has links to a number of interesting articles on dissection.

Index

absolute, 98
All-Star Tessellator, 130,139
altitude, 172, 211
area warm-up, 172

base ten representation, 64, 74
Bolyai-Gerwien Theorem, 225
boundary of a polygon, 178
    of a polygon

convex polygon, see polygon

density of a packing, 147
diameter of one penny, 144
digit
    dexterity, 62
    units, 62, 64
    tens, 62, 64
    hundreds, 74
dissection, 202
distributive law, 13
divisor of a number, 20
duality, 42, 44, 49
    double-dual square, 50
    double-dual triangle, 44
    dual square, 50
    dual triangle, 43
    dualizing process 43

equilateral triangle, 26, 186, 197
    basic, 159

factorization of a number, 20
flip (reflection), 30, 37, 41, 48, 95

game
    Four Numbers Game, 78, 98
    Pick's Polygon Game, 183
    Race to 100, 4
    Roll Back, 10
    Square Game, 34
    Tell-Me-If-It-Tessellates Game, 115
    Triangle Game, 27
    Two-Digit Palindrome Game, 62
    Wordsworth, 16
geometric constructions (review), 222

hexagon, 132, 137–139
    special, 123, 125, 126
    regular, 139
hexagonal packing, see packing
house pentagon, see pentagon
hundreds digit, see digit

Jordan, Michael, 16, 18

k-step palindrome, 60
kite, 122

lattice, 172
    point, 172
    polygon, 170
length of The Four Numbers Game,
    80

Major League Tessellator, 127
median of a triangle, 32
monster-gon, 131

n factorial, 22
nasty-gon, 131, 175

non-convex polygon, see polygon
number line, 86

packing 144
    hexagonal, 146
    square, 146
    repeat triangle, 152
    triangle, 146
    tight repeat triangle, 167
palindrome, 54
    five-digit, 56
    four-digit, 56
    zone, 64, 70
    table of k-step two-digit, 71
    three-digit, 55
    two-digit, 54
parallelogram, 211, 214
    extreme in-case, 229
    in-case, 227
    out-case, 227
parity, 104
pentagon, 132–136
    house, 134
    regular, 133, 135
perfect square, 188
Pick
    Pick's Polygon Game, 183
    Pick's Theorem, 183
polygon, 131
    convex, 131, 237
    lattice, 170
    non-convex, 131, 175, 237
    review, 130
    regular, 131, 200
    triangulation of, 173, 237
prime number, 21, 76
Pythagorean Theorem, 180, 191, 218,
    243

quadrilateral, 121, 127
    with supplementary adjacent an-
    gles, 127

reflection (flip), 30, 37, 41, 48, 95
repeat triangle packing, see packing

reverse and add rule, 58
rhombus, 119
Rookie Tessellator, 119
rotation, 29, 36, 41, 48, 94, 118

scalene triangle, 119
side sum, 27
skew lattice
    line segment, 190
skew lattice
    square, 173
square packing, see packing
standard form of The Four Numbers
    Game, 98
start square, 78
step number, 65, 72
subtraction rule (routine), 78, 84
symmetry property of dissection, 205

tangrams, 206
tens digit, see digit
tessellation (tiling), 114
    by pure translation, 117
    tessellating shape, 115
    vertex-to-vertex and edge-to-edge,
        114
Tessellators Hall of Fame, 142
tiling, see tesselation
triangle packing, see packing
triangulated polygon, see polygon
Tribonacci numbers, 109
transitive property of dissection, 239
type (class) of tessellating convex hexagon,
    137–139
type (class) of tessellating convex pen-
    tagon, 135, 139

units digit, see digit

values of letters of the alphabet, 16
vertex-to-vertex and edge-to-edge tiling,
    see tessellation

Well Ordering Principle, 105
Williams, Serena, 16

zero game, 82